# Landscape Agroecology

# Landscape Agroecology

Paul A. Wojtkowski

Routledge
Taylor & Francis Group
New York  London

First published 2004 by Haworth Press, Inc

Published 2018 by Routledge
52 Vanderbilt Avenue, New York, NY 10017
2 Park Square Milton Park, Abingdon Oxon OX14 4RN

*Routledge is an imprint of the Taylor & Francis Group, an informa business*

Cover design by Marylouise E. Doyle.

**Library of Congress Cataloging-in-Publication Data**

Wojtkowski, Paul A. (Paul Anthony), 1947-
    Landscape agroecology / Paul A. Wojtkowski.
        p.  cm.
    Includes bibliographical references and index.
    ISBN 1-56022-252-2 (hard : alk. paper) — ISBN 1-56022-253-0 (soft : alk. paper)
    1. Agricultural ecology. 2. Landscape ecology. I. Title.
S589.7.W648 2003
577.5'5—dc21

                                                                                        2003006841


ISBN 13: 9781560222521 (hbk)
ISBN 13: 9781560222538 (pbk)

Then We cause to grow thereby gardens of palm trees and grapes for you; you have in them many fruits and from them do you eat;

The Koran, The Believers 23.19

. . . and He causeth to sprout from the ground every tree desirable for appearance, and good for food, . . .

The Bible, Genesis 2:9

# ABOUT THE AUTHOR

With over 25 years of experience in many corners of the world, **Paul A. Wojtkowski, PhD,** is uniquely qualified to deal with landscape agroecology. Dr. Wojtkowski has resided in eight countries and has conducted professional observations in another seventy. Some of his residences were in environmental trouble spots or in regions where food and fiber are grown only through nature-accommodating measures. The list includes Africa (almost four years), Australia (one year), the South Pacific (two years), Latin America (three years), Europe, and North America.

From these travels, Dr. Wojtkowski offers insight into how cultures and peoples put their imprint upon the land. In presenting this material, he is also able to couple field-level practical experience and an applied perspective with a strong understanding of theoretical underpinnings. This is Dr. Wojtkowski's fourth book. His first two are in agroforestry; the third addresses general agroecology.

# CONTENTS

# Preface

Agroecological landscapes are those that have productive purposes. They can involve agriculture and/or forestry, and be intensely managed or lightly touched, but still human-influenced, natural ecosystems (e.g., a logged forest). The effects of a well-managed, environmentally friendly landscape extend outside an immediate area. Degraded terrestrial ecosystems can, through negative association, directly affect nearby, or not so nearby, ocean ecosystems.

A fully agroecological landscape would be more in harmony with, and would often use, natural processes. A more agreeable end goal is a landscape formulated using ecological dynamics to back a productive role and as a by-product maintaining, through decorous use, respect for the land, the natural processes, the vegetation, and the living creatures therein.

## Ecology versus Agroecology

Implicit in this text is the idea that agricultural ecology, rather than natural ecology, is a more suitable vehicle to explain and advance human-influenced landscapes. Both natural and agricultural ecology are branches of ecology and share a host of commonalities. However, these two disciplines view natural processes from different perspectives, have divergent end goals, and proceed along some distinct lines in regard to the theories, principles, and practices espoused.

Part of this lies in the desired harvests, yield levels, and economic objectives that, in untouched, nonimpacted, and nonmanaged ecosystems are entirely lacking. Furthermore, a whole string of landscape structures (fields, orchards, hedgerows, dwellings, barns, roads, along with ownership patterns) promote the productive use of the rural countryside. These also are not part of natural ecology.

Other human needs and values also incline a landscape to one form or another. This, coupled with an academically evolving set of agroecological principles, is what underlies the agroecological landscape. Outside the bounds of agroecology, but within the sphere of natural ecology, is the study and monitoring (as opposed to the purposeful management) of natural ecosystems.

## *Basic Agroecology*

Plant-on-plant ecology lies at the core of any discussion on agroecology. The ramifications of the plant-plant interface are the focus of other texts and are only touched upon here. The plant-plant interface can be a starting point that ultimately ends at regional (and landscape) agroecology.

Although plant-on-plant and plot agroecology are important and cannot be separated from the whole, development in this text centers more on the one-plot, one-agroecosystem land-use model. The discussion expands along a number of paths—spatial, temporal, and socioeconomic—eventually discarding the one-plot one-agroecosystem model in favor of other, often more culturally influenced, alternatives.

It is hoped that the material covered approaches and is representative of a culturally and climatically diverse world. Travel and study in all regions is not possible, and a scattered and fragmented literature base does not guarantee that the full scope of agroecological achievement is included.

## *Biocomplexity*

As biocomplexity grows, so does the number of possibilities. An interesting parallel in mathematics illustrates this point. The multiplication of two numbers does not offer much methodological variation. In linear algebra, there are a number of techniques to multiply number arrays, each being useful, each expanding the horizons of both theoretical and applied mathematics.

In agroecology, the simplest unit is a single crop growing on a single sequestered plot. As biocomplexity grows, so do the options and applications. Although sorting through the array of options can be perplexing, among the multitude are many that exceed in value (in terms of both output and environmental advantage) those gains that accrue from a landscape composed of simple, unrelated, and noninteracting monocultures.

This book attempts to simplify what can be a very complex topic, not by restricting what is covered, but by providing routes through this intricacy. In this way, it is hoped that the challenge posed by the complexity will be less daunting, and meaningful solutions can be reached.

As with the choices gained through the mathematics of arrays, many approaches and concepts underlie agroecology. In an agroecosystem-based approach, these include complementarity, desirable plant characteristics, competitive production, facilitation, etc. Each provides a level of insight and a path into the complexity of multi-species, multiagroecosystem agriculture and forestry.

Similarly, landscape agroecology offers some singular viewpoints, a host of concepts and theories, along with an approach that is complementary with agroecosystem ecology. If any drawback exists, it is that some of the ideas and concepts must be dismantled and rebuilt in a somewhat different, but not alien, form.

### *Cultural Agroecology*

There is much to be learned about how human cultures interact with the land. Given the amount of variation and the potential ecological gains from many unique and unexamined land-use practices, this affords a rich avenue of study. The gains are realized by understanding the underlying agroecological dynamics, the cultural and socioeconomic motives, and the site conditions that bring unique practices to the fore.

This text fosters the idea that, within different societies and cultures, ancient or modern, developed or subsistence, superior land-use techniques have evolved. Study is best accomplished by offering a framework by which knowledgeable observers can differentiate the unique from the mundane.

### *Basic Outline*

This text is divided into 14 chapters, with the following topics:

1. Introduction
2. Basic agroecological concepts
3. Understanding the agrotechnologies
4. Principal-mode agrotechnologies
5. Temporal and auxiliary agrotechnologies

6. Water management
7. Wind, frost, and fire management
8. Integrated pest management
9. Physical and temporal patterns
10. Landscape socioeconomics
11. Biodiversity
12. Other landscape influences
13. Cultural motifs
14. Conclusion

At the end of Chapters 6, 8, and 10 are some brief case studies. Although these highlight particular topics, they also show that each topic is only one part of a broader picture and part of the cross-effects inherent in an agroecological landscape.

## Opinions Expressed

The concepts behind an agroecologically designed landscape are, in academic terms, comparatively new. From scattered studies and everyday observation, it is possible to paint a series of pictures of ecological and human diversity within various landscape layouts.

Some of the topics, e.g., modeling, landscape design approaches, and cultural motifs, may seem abstract and less relevant. One must keep in mind that these are part of the decision process. Despite not being explicit in current literature, they help explain disparities in viewpoint and land-use practice.

As with many disciplines, the topics are not linear progressions and do not easily fit such a form (i.e., from Chapters 1 to 14). Landscape agroecology is a series of overlapping topics where the presentational ordering constitutes yet another view.

As with any new discipline, the views and approaches are a work in progress. The material presented here should be looked at in the light of provoking debate and fostering thought, with the purpose of providing a greater understanding and appreciation of the rural countryside. An idyllic society is well outside the province of agroecology, but a Garden of Eden is more realizable than many may think.

# Chapter 1

# Introduction

Landscape agroecology, the ecology of a productive countryside, is a branch of general agroecology. It is also an offshoot of agriculture, forestry, agroforestry, and natural ecology, where the focus is not on individual (agro)ecosystems (agroecosystems and on-farm natural ecosystems), but on (1) the interaction between human-derived, -managed, and/or -influenced ecosystems or (2) the interaction between those that yield useful output and neighboring natural and nonproducing ecosystems.

The goals of agroecology are primarily productive, but, with the ecological underpinnings, the emphasis and resulting methodologies do differ from more mainstream approaches to agriculture and forestry. In a moderate form, an agroecologically influenced landscape is an environmentally friendly, productive expanse where the land-use patterns assert and reinforce the socioeconomic and cultural goals of the inhabitants.

A stronger statement of purpose, one that rests firmly upon ecological underpinnings, is that landscape agroecology is designed to use ecological dynamics to promote or achieve productive purposes and the betterment of the human experience. Part of this is developing a landscape that allows natural flora and fauna to thrive in minimal competition with the productive role of the land.

The productive outputs are those associated with traditional agronomy and forestry. Normally, these are obtained with planned and managed agroecosystems. Outputs can also be obtained through the sustained exploitation of naturally occurring flora and fauna.

The human experience can be harder to grade. The wide-ranging benefits include economic returns, quality-of-life gains (clean potable water, beautiful scenery, etc.), or hunting and gathering in untamed natural ecosystems. In short, landscape agroecology covers all forms of human consumption obtained from terrestrial ecosystems and, by default, includes many land-enclosed and/or land-affected aquatic systems.

## LAND-USE PATTERNS

The figures in Table 1.1, compiled by region and for selected countries, give some insight into the land area managed for human purposes, what is available, and what is left untouched.

Through Table 1.1, it is clear that areas under permanent cultivation may be less than travel and simple observation suggest. This may be because cultivated land is highly visible, often located along roads and other transportation links. More telling is that populous countries, such as India and China, have comparatively small percentages of their arable land in permanent cultivation. This and other country data imply that land area exists to expand cultivation. This varies by country and, in some cases, may require the use of more marginal land.

### Human Impact

Having underutilized land does not mean that human impact is limited. The data (Table 1.1) most likely understates the effect of humans on natural terrestrial ecosystems, protected areas included. Most human activity is concentrated in higher-fertility zones, impacting some ecosystems more than others. Also, a fair percentage of noncropped arable and protected areas experience human encroachment, through timber harvests, hunting, gathering, and/or grazing.

There is a consensus that the growing population of the earth is putting stress on all terrestrial ecosystems and, directly or indirectly, through poor land practices in coastal regions, on far-ranging non-terrestrial resources such as ocean fish (e.g., Kühlmann, 1988). This may be through urban expansion, poor land management, pollution, climate change, or other long-term, less observable means.

TABLE 1.1. The total land area, arable land, land under permanent cultivation, and protected areas for regions and selected countries

| Continent/country | Total land | Arable land | (%) | Permanent crops | (%) | Protected area | (%) |
|---|---|---|---|---|---|---|---|
| Africa | 2,966,876 | 174,907 | (6) | 24,431 | (1) | 154,073 | (5) |
| Egypt | 99,544 | 2,834 | (3) | 466 | (<1) | 7,864 | (8) |
| South Africa | 122,104 | 153,360 | (12) | 940 | (<1) | 7,143 | (5) |
| Kenya | 56,914 | 4,000 | (7) | 520 | (1) | 3,420 | (6) |
| North America | 2,134,950 | 237,374 | (11) | 7,885 | (<1) | 183,684 | (8) |
| United States | 915,912 | 176,950 | (19) | 2,050 | (<1) | 101,849 | (11) |
| Mexico | 190,869 | 2,520 | (1) | 2,100 | (1) | 248 | (<1) |
| Costa Rica | 5,106 | 225 | (4) | 280 | (5) | 640 | (12) |
| South America | 1,751,708 | 96,004 | (5) | 20,597 | (1) | 1,059 | (6) |
| Argentina | 273,669 | 25,000 | (9) | 2,200 | (<1) | 4,296 | (1) |
| Brazil | 845,651 | 53,300 | (6) | 12,000 | (1) | 31,965 | (4) |
| Chile | 74,880 | 1,982 | (3) | 315 | (<1) | 13,673 | (18) |
| Asia | 3,088,370 | 498,849 | (16) | 58,503 | (2) | 137,641 | (4) |
| India | 297,319 | 161,950 | (54) | 7,900 | (3) | 13,468 | (5) |
| China | 932,641 | 124,145 | (13) | 11,220 | (1) | 56,424 | (6) |
| Japan | 37,652 | 3,915 | (10) | 380 | (1) | 2,808 | (7) |
| Indonesia | 181,157 | 17,941 | (10) | 13,046 | (7) | 17,517 | (10) |
| Europe | 2,236,976 | 293,335 | (13) | 18,249 | (<1) | 119,074 | (5) |
| United Kingdom | 24,160 | 6,380 | (26) | 45 | (<1) | 5,059 | (20) |
| Germany | 34,927 | 11,832 | (33) | 228 | (<1) | 9,000 | (25) |
| France | 55,010 | 18,305 | (33) | 1,163 | (2) | 5,666 | (10) |
| Oceania | 846,769 | 54,739 | (6) | 2,669 | (<1) | 99,885 | (11) |
| Australia | 768,230 | 52,875 | (7) | 225 | (<1) | 93,570 | (12) |
| New Zealand | 26,799 | 1,555 | (6) | 1,725 | (6) | 6,214 | (23) |
| Papua New Guinea | 45,286 | 60 | (<1) | 610 | (1) | 81 | (<1) |

*Source:* UN, 2000, p. 641.
*Note:* Area in thousands of hectares; percentage of total land area.

In a few situations, conservation and agroecology are less or not applicable, such as (modified from Kumar, 1999)

1. where new lands are available for exploitation and utilized and spent areas can be abandoned for regeneration,
2. where new resources replace the old (e.g., where annual flooding replaces lost nutrients), or
3. where cooperative behavior is lacking (lawlessness or unmanaged common property) and no person benefits from conservation efforts (as with unregulated common lands).

With high population levels and the need to draw upon a range of resources, there is currently little occasion to ignore sustainability concerns. The exceptions are few and do not apply beyond some scattered local, temporary, or narrowly interpreted situations.

The forces that cause food and environmental problems are more complex than increased population densities. Some blame may be attributed to regional climatic fluctuations. An example is a change in local rainfall patterns brought about by agricultural expansion and the loss of natural land cover.

Market forces and the need for cash crops can compel change and outstrip the ability of local communities to respond using their knowledge of conservation practices (Henrich, 1997). This situation can occur when local land use systems and their biodiversity are sacrificed for a narrow range of productive outputs. In a worst-case scenario, social structures may loose coherency under resource pressure, precipitating or further aggravating a bad situation.

Except for social tumult, there are few circumstances where, through the application of agroecological principles, a worsening land use situation cannot be reversed. This is not always through more inputs (labor, chemical fertilizers, herbicides, etc.), but can occur through the more sophisticated use of local resources, more accommodating agroecosystems, and a better agroecological landscape.

### Agroecological Need

In the terrestrial sphere, sufficient land area is available to feed the earth's population with some margin to spare. There is also ample opportunity to accommodate increased production, to reduce risk, and to do so with an eye toward preserving natural ecosystems and their flora and fauna.

Production can be population dependent in a positive direction, where productivity can and does improve with population growth (Pearce, 2001). Mingkui Cao et al. (1995) projected that, barring unexpected occurrences, the agricultural resources of a populous country, such as China, can support 51 percent more inhabitants. These gains, both productive and environmental, require some preconditions:

1. Land-users with a long-term outlook and the land control to realize it
2. An ability to change agricultural or forestry practices in response to environmental need (including the new knowledge required)
3. Respect for nature that can resist some economic pressure
4. A strong and stable social order

In regions where these conditions exist, such as North America and Europe, food and fiber productivity has generally surpassed population growth. The technological change that predicated these gains offers a further opportunity to introduce cropping systems that are highly productive, low input, and environmentally friendly. The land freed by concentrating production in a smaller area also offers possibilities for beneficial land use change that further accentuates the positive gains across the larger landscape.

In some regions, the effect may not be strong, but the results are equally viable. For example, in Kenya, tree planting by farmers exceeds population growth (Holmgren et al., 1994). The loss of productive capacity through overuse or mismanagement encourages (or forces) people to adopt new, more conservation-oriented practices. The same has been noted in the west African Sahel (Glausiusz, 2003). Again, positive shifts at a plot level, coupled with understanding of local ecology and change, can further drive overall land use practice in a positive direction.

History supports the premise that through the application of appropriate practices, marginal lands or unorthodox agricultural and/or forestry settings can be high yielding. During Roman times, parts of arid North Africa yielded much more wheat than is currently possible.

This was due to impressive irrigation systems that provide little water today (Hughes, 1994, p. 142) and local areawide conservation measures that have fallen into disuse (e.g., Nevo, 1991). In other regions, farmers still use ancient practices that are appropriate, capable of being updated, and that can be brought within mainstream agroecological thought. Current agroecological texts (e.g., Gliessman, 1998; Wojtkowski, 2002) contain numerous examples of local practices with wider potential.

Through increased per area productivity, options open for conservation efforts over wide areas, which offer a general improvement in productivity, sustainability, and quality-of-life factors. By providing the necessities as well as quality-of-life benefits, locals have more opportunity to acquire or increase their respect for nature. With this can come the capacity for further gains in the productive and ecological value of the land.

## THE FIELD OF LANDSCAPE AGROECOLOGY

As a field of study, landscape agroecology begins with plots and agroecosystems and expands through the principles and practices of plant-on-plant agroecology. The goals of these two versions of agroecology are the same, a productive and environmentally sound countryside. Through the two versions, more options can be directed and coordinated toward the same tasks and end goals.

In some aspects, landscape agroecology is akin to a free-form jigsaw puzzle where the color of the pieces (land units) corresponds to different land uses. The user is free to derive the best panorama using some or all of the pieces provided. These pieces are the existing plots or agroecosystems with the option to use introduced systems or vegetative additions to promote various ecological effects.

Beyond and within the physical landscape, there are many competing influences and underlying concepts. The best results come from informed decisions that take into consideration as many options as possible. To achieve this, this book examines competing ecological, agroecological, socioeconomic, and cultural influences. These form a basis for informed decisions that span and incorporate the views and aspirations of those involved. This chapter begins the process by looking at some of the fundamental units of the physical landscape.

## Ecosystems

An agroecological landscape is composed of various ecosystem types: those which are fully planned and managed, those with a large natural component, and those whose dynamics are dictated entirely by natural forces. A landscape can be all of one type or, more likely, a mix of (agro)ecosystems. How the individual systems are subdivided, used, and placed is a focus of landscape agroecology.

### Planned Agroecosystems

Planned systems are based entirely on the interactions between living components purposely put on a given land area. Although all components are planned, these resulting interactions must be at least partially intentional. The added flora or fauna exist because of planning and management, and their presence contributes to achieving set land use objectives.

The area can be as small as a single plant (e.g., a large, prominent tree) or include the vegetation covering a large tract. The only stipulation is that an agroecosystem exert enough influence, whether agronomic, economic, or ecological, to be a measurable force in the larger landscape and a significant influence in achieving one or more land use goals.

### Occurring Agroecosystems

In planned ecosystems, unplanned encroachments often occur. Unplanned incursions can include weeds, herbivore insects, unintended fauna, etc. These often exert a negative influence with regard to system objectives, although positive interactions can occur.

The occurring agroecosystem is the result of continuing natural forces that attempt to convert a planned ecosystem to a natural ecosystem. Weeds are the most common example, although a bird population, if provided with food sources and nesting sites, can be an unplanned and possibly unwelcome intrusion.

Many forestry plantations and orchards, although designed as single-species systems, contain considerable unplanned biodiversity. Most of these incursions are relatively benign, but if the single or combined results become adverse, management input can restore the status quo.

*Natural Ecosystems*

Natural ecosystems are those interactions that are outside the context of any land management scheme. In essence, this is what nature does with an area when human management does not exist. Such areas lack planning and management, but are part of agroecology on three fronts:

1. Where hunting, gathering, and/or tourism is an economic activity not requiring any ecosystem change
2. Where they are part of overall landscape (agro)ecosystems (as with forest borders or fragments)
3. Where they serve to guide an emulation or mimicry approach to agroecology development

In the first case, natural ecosystems are a large component of a landscape. Land users obtain economic benefits by utilizing these resources. Logging in natural forests is a common example.

Natural ecosystems are an entity within a landscape that should, with proper management, never be a negative influence. As such, the size and placement of these systems is a useful tool in achieving wider agroecological objectives. Their borders, with the spillover of ecological dynamics, can be harnessed to provide good.

The natural ecosystem is the result of many influences (soil type, climate, disturbance, etc.) and subdivides into vegetative groupings in response to these influences. This is of interest in agroecology, as the natural ecosystem, both in the mature and/or immature phases, provides insight into the types of plants and agroecosystems best suited to a site. This is the idea behind agroecological mimicry, in which biorich agroecosystems duplicate the dynamics of natural ecosystems.

## CONCEPTS OF LAND USE

### *Land-User Units*

In addition to classifiable groupings of vegetation and observable natural features, human prejudices subdivide large tracts of land. These divisions can follow natural influences or features or serve only the perceived needs of land users. The types of subdivisions include holdings and economic areas.

### *Holdings*

A holding is a land area where a person, family, or small group has obtained legal, traditional, or other use rights. In the absence of formal documents, this can constitute a claim or land privilege that is recognized by a local population. Ownership, even in its strongest form, may involve certain infringements whereby some land use prerogatives are taken away or reduced. Zoning or mineral rights are examples, where the full scope of land use activities by the owner are restricted.

The holdings of most interest in landscape agroecology are commonly called farms or plantations, the latter being used in forestry (destined for harvest and conversion to some wood product) or with tree crops (orchards and other perennial plantings with a nonwood output as the primary goal). The farm or plantation can be the distinct unit within the larger regional landscape and, as a productive entity, may transgress other smaller contained units (as with sharecrop divisions).

Having said this, rare situations exist where farm or plantation rights may be superior to legal boundaries. Informal grazing or other traditional rights and practices can often cross legal and, at times, even stronger national boundaries (e.g., Meir and Tsoar, 1996).

### *Economic Area*

The economic area is the amount of land area subject to measurable economic activity by a single economic entity (person, family, corporation, or other definable group). Measurable economic activity involves the production of goods and services within the subscribed holding. A holding can be divided into an unused and unimpeded natural area and a farm sector. The farm sector is the economic area.

An economic area can extend outside of a legal area (holding) where, e.g., land on a neighboring farm is rented. Community agriculture or forestry may have an economic area where crops or trees raised across legal boundaries, can encompass a wide area and have shared rights and productive responsibilities.

At the lower end of the economic spectrum, and often outside the legal boundaries, is subsistence hunting and gathering in a minimally disturbed natural ecosystem. If this occurs outside a holding, the economic area is larger than the holding.

For example, Robinson Crusoe, alone on an island, was an economic entity, the used part of the island was the economic area or unit, and, in the absence of other exercised legal or traditional rights, the island was also a holding. His economic activity was hunting and gathering.

Other examples of economic entities are an extended family where relationships link farms, or a company, corporation, trust, or cooperative. There is no reason why economic activity cannot be shared among different entities. Subsistence gathers can have overlapping economic areas, often enshrined with complex and informal land rights.

### Agroecological Units

A rural landscape can be divided into observable and definable subunits. These come in different forms and subdivide the landscape into smaller areas based on agroecosystem age, locational convenience, economic need, management, or land use compatibility. This is fairly abstract and includes two types of productive units, plots and agroecosystems, which can be one and the same or quite different. The plot and agroecosystem differ in that one (the plot) is basically management oriented and the second (the agroecosystem) has groups of plants in ecological agreement for a specific purpose.

Other agroecological units either span or are subsets of other units (e.g., a composition unit). They have various functions within the landscape. The differences are subtle and can defy observation and simple definition.

## Productive Units

The plot containing a single agroecosystem (the one-plot one-eco-system model) is often the most prominent feature in many, but not all, rural landscapes. Some cultures have proceeded in other directions, spurning clearly defined agroecological units, either plots or agroecosystems, in favor of a more free-form approach.

Although agroecosystems do not often transcend clearly defined plots, this does not hold true for more loosely demarcated plots. A plot can contain a series of agroecosystems bound by a common purpose. A legal entity may not be divided into plots, but may contain interlocked and overlapping ecosystems.

Transcending these is the capability to arrange plots and/or agro-ecosystems within a landscape to achieve both narrow and broader objectives. This section describes these units. More detailed explanations follow in subsequent chapters.

*Plots.* Plots are land areas often clearly demarcated by boundaries either introduced (e.g., fences and hedges) or natural (e.g., streams, steep hillsides, etc.). These often separate plants grouped by species, age, economic need, type of management, or some other criteria. A plot can be composed of a single agroecosystem or a collection of agroecosystems that serve a common objective within the limits of the demarcated area.

The plot can be more of a management convenience than a pure productive unit. In this case, the plot serves as land use guide and as a convenient partition within a larger farm or forestry entity. Normally, plots enclose one or more agroecosystems (as shown in Photo 1.1).

*Agroecosystems.* The agroecosystem is the central agroecological unit within any landscape and, although agroecosystems need not contribute directly to landscape productivity, this is also the main productive unit. These can be contained within or extend beyond marked plots and can be composed of a single species, mixed species, or integrated subunits.

What defines the agroecosystem is the design role and planned resource complementarity between the component plants and/or species. All subunits in an agroecosystem are dedicated to achieving one set of objectives (e.g., production, erosion control, etc.). A requirement is a high degree of planned plant-plant complementarity (as detailed in Chapter 2), such that the objectives are reinforced.

PHOTO 1.1. This single plot encloses two distinct, but functionally overlapping agroecosystems. Within the pasture is a forestry (pine) plantation with grazing as a secondary purpose.

   Monocultures are agroecosystems because these species share a common purpose and site resources (i.e., based on the spatial allocation of resources between niche similar plants). Intercrops and multispecies plantations have more complex resource strategies that can involve the capture, allocation, and/or flow of resources between component species. With polycultures, resource intent is a key factor in designating unlike plants as part of a single agroecosystem.

   *Management/Composition Units.* Agroecosystems and/or plots can be subdivided by the type, age, and/or management of the vegetation, or by site characteristics. The difference can be apparent but, in some situations, determination may require more than visual observation. For example, forest gardens are often subdivided by the vegetation mix and output focus of each subsection (e.g., Gamero et al., 1996).

   In more unusual cases, a farm or tree plantation can be in essence a single plot and one agroecosystem with different composition units (e.g., age). For example, in Bahia, Brazil, there are cocoa farms larger than 200 ha that are subdivided into age groups. When the rotations are very long (greater than 100 years), these in essence become a single subdivided agroecosystem. These subunits, although difficult to distinguish, may still exist and form the basis for management decisions. This can also occur with large tree plantations (as shown in Photo 1.2), with large tree crop holdings (e.g., rubber tree, coconut, or oil palm plantations), or in large orchards.

## Specific Interaction Zones

Specific agroecological interactions are those that overlap ecosystems, involve flora and/or fauna, and consist of unique ecological interactions between plots, agroecosystems, and/or larger landscape units. An example of such interactions may be between hedgerow and crop area. Predator-prey relationships may originate in the hedge and overlap partially into the crop plot. This is one interaction area. A second may involve windbreak effects. Again, this begins at the

PHOTO 1.2. A mixed grazing-pine plantation in New Zealand. This system covers most of the holding and can be subdivided into managed plots based upon management convenience (e.g., the pruning evident in the photo).

hedge and extends into the crop plot, where the second ecological effect within this zone may not coincide in area and influence with the first.

A specific interaction zone (SIZ) has three components: magnitude, size, and duration. Magnitude is a measure of how strong the effect is and how effectively it accomplishes the economic and/or environmental task. Size refers to how much land area is covered or affected. Duration is the temporal component. Some effects are permanent, others temporary.

## Macroecological Area

The largest of the agroecological areas, macroecological area, is extraterritorial, i.e., outside the control of individuals or groups of land users. There are a number of cases where a macroecological area has influence over smaller farm or forestry units. For example, deforestation can alter the climate and/or reduce rainfall over a wide area. Yoon (2001) reports reductions in moisture in the cloud forests of Costa Rica due to unadvised agricultural activity. Major insect outbreaks can also extend over a wide area.

Other macroecological areas are smaller. For example, the water dynamics in a large watershed constitute a macroecological influence; on a hillside, they have a smaller macroecological domain; and for a specific agroecosystem or specific flora and/or fauna, the area is reduced to an SIZ.

# PERSPECTIVES ON AGROECOLOGY

How agroecology is viewed can be all-important. Most of the research has looked at various productive units. This text takes the perspective of the larger landscape, based on the use and placement of the smaller components. Given the complexity of the topic, this does not exhaust the list of approaches; other perspectives exist.

## Plant-Plant Agroecology

Ecology is the study of organisms and their environment. The smallest unit is a single plant. Since agroecology involves productivity in planned and managed ecosystems, the smallest practical study unit involves the interaction of two plants grown in close proximity. These can be niche identical (as with clones) or niche dissimilar (as with different species).

## Complex Biodiversity

It is equally viable that agroecology and/or a landscape can be based on species-rich, complex ecosystems and the natural dynamics inherent in these systems. These are landscapes that transcend plant-plant dynamics and instead are formed around dynamics of complex ecosystems. This encompasses the concept of mimicry where agriculture or forestry seeks to replace what nature does with a landscape. Instead of realizing a natural ecosystem, these dynamics, through species and their placement, are directed toward productive purposes.

## Economic Gains

Not to be overlooked are the economic requirements of agroecology. Farming and forestry want to achieve maximum output with minimal input, and this can force land users down certain roads. If agroecology can offer the same or a better mix of inputs and outputs, this can overcome much of the reluctance to adopt a more agroecological, environmentally sensitive approach, knowledge of the options permitting. When agroecology is less economically attractive, this potential must be developed or other directions pursued.

## Cultural Parameters

In an agroecological landscape, the basic unit is often the agroecosystem. Landscape agroecology looks at the interrelationship of these basic units and how to design to achieve the best possible outcome. The view taken stems from training and experience and is tempered with a dose of culture. Holdings show many cultural manifestations, e.g., cattle and pigs have a religious taboo in some parts of the world. In other areas, these are an important element in the diet and land use practices provide for them.

Cultural influences are part of wider human-nature tenets. These are advanced in ensuing chapters. Cultural influence is the idea that there are some underlying thoughts, views, or perceptions that ultimately govern the landscape design process. They translate, through motifs, into an active landscape. The landscape motif is the pattern or theme that allows groups and/or individuals to work within their knowledge base without being overwhelmed by options or alternatives.

## Landscape Motif

Motif many be the most abstract of the landscape concepts, because the human-nature interface (how people view and utilize natural and modified ecosystems), in its applied form, may be manifested only through land use practices.

Within a region, landscapes are similar because land users share these characteristics:

1. Similar site and climatic situations
2. Common socioeconomic circumstances (including land holding size)
3. Climatically focused dietary needs (staple crops)
4. The same degree of risk
5. A similar background with land use issues

The point is that different cultures and peoples do not subscribe to the same views on agroecology, and model the landscape on their needs and perceptions. Understanding this aids in studying diverse agroecosystems and in extracting the full value from any landscape.

# Chapter 2

# Underlying Agroecological Concepts

The placement and use of plants and ecosystems is a key element in landscape agroecology. To fully understand the inherent dynamics requires understanding a series of underlying concepts. A large share of landscape dynamics starts with the individual plants, and some of the ideas that underlie individual productive units, either the plant or ecosystem, have application on a larger scale. Other concepts transcend individual productive units and are a force only in the overall landscape.

## ESSENTIAL RESOURCE MANAGEMENT

One aspect in any agroecological landscape is the need to manage essential resources, including light, water, $CO_2$, the primary nutrients (N, P, and K), and a host of trace elements that aid in plant growth (S, Ca, Fe, etc.). Some resources are controlled through internal ecosystem dynamics, while others, mainly through facilitative effects, can cross ecosystem lines.

### Light and Water

As essential resources, light and water are transitory, not lasting long within an ecosystem and requiring continual resupply. Although there are only a few cases where light is an interplot effect, there are means to ensure the efficient use of water. A landscape can be used to supply water, as a means to conserve moisture as a scarce resource, to manage when in oversupply, and to control unwanted side effects. This subject is addressed further in Chapter 6.

## Nutrients

There are three options for capturing and cycling nutrients within agroecosystems: (1) simultaneous capture and use, e.g., perennial systems; (2) sequential capture and use, e.g., with a fallow cycle; and (3) supplemental inputs. The first two use plants in a facilitatory role, both within and between plots. The third uses manures and chemical inputs or involves cross-ecosystem transfer (e.g., green manure).

## COMPLEMENTARITY

In Chapter 1, agroecosystems are spatially defined by the intended degree of plant-plant complementarity. This is based on the four fundamental agrobionomic principles: (1) competitive acquisition, (2) competitive partitioning, (3) facilitation, and (4) competitive exclusion (Wojtkowski, 2002, p. 34).

The first of these, competitive acquisition, means that, through the niche diversity of intercropping, unlike plants can capture a greater amount of available essential resources, which can give overall productive gains compared to monocultures lacking niche diversity. The second principle, competitive partitioning, occurs in resource-rich situations where, through biodiversity, more efficient use of essential resources and greater overall productivity results.

These principles confer essential resource compatibility and have application on a number of levels. The base unit, plant-on-plant interactions, can be

1. distinguished economically through competition (or the lack thereof),
2. used to ecologically define agroecosystems within the physical landscape, and
3. used in determining the agronomic or biological efficiency of individual productive units or an overall landscape.

The third agrobionomic principle, facilitation, means that the presence of one plant species benefits another, which results in productivity gains. As essential resources can be moved or directed between landscape units, facilitation can transcend these units. This can be a governing or important design influence in the overall landscape (see Photo 2.1).

PHOTO 2.1. Oil palms with understory. This triculture has facilitative non-productive hedges with center rows of pineapple. On this site, the high degree of interspecies interface presupposes a high degree of interspecies complementarity.

The fourth principle, competitive exclusion, states that by eliminating or suppressing competition, more essential resources are available to the more desirable species. Although useful within individual agroecosystems, exclusion has little application at the landscape level.

Clearly, some plant-plant combinations achieve complementarity through one or two niche differences. For example, in China, the paulownia *(Paulownia elongata)* achieves complementarity with crops in part due to root strata and temporal separation. On the west African savanna, the tree species *Faidherbia albida* relies on temporal separation. In contrast, the complementarity of the widely used maize-bean association is harder to pinpoint and may be due to broad efficiency gains (competitive partitioning and facilitation).

## Measurement

Complementarity is defined through the land equivalent ratio (LER). With LER, values above one indicate that two separate species can achieve greater productivity when grown in close proximity than plants in an equivalent monoculture. Although measurement is straightforward (see equation 3.1) and much can be ascertained, the LER does not separate the agrobioeconomic principles and provides little insight into how to best use complementarity and how to maximize this value.

## Subdivisions

The basic idea is that an agroecosystem (or plot with the one-plot one-agroecosystem model) is defined by plant-plant complementarity between groups or individual plants. Normally, an LER equal to 1 (by definition for monocultures) or greater than 1 (for polycultures) is expected. Exceptions do occur, and some agroecosystems may be formulated on partial or temporal complementarity in a highly integrated intercrop. Generally, if an LER greater than 1 is not expected or anticipated (it may happen by chance, but this does not matter), then the vegetative groups are separate systems.

To expand this through example, if a plot is bounded on four sides by hedges and these are not anticipated to be complementary with an internal planting regime (i.e., the agroecosystem), then the hedge and internal agroecosystem are separate landscape units (i.e., the hedge is not part of the enveloped agroecosystem, but is a separate boundary system). If a hedge species is selected to have plant-plant complementarity, then it and the internal plants are a single agroecosystem.

Monocultures, by definition, have an LER of 1 and can be separate agroecological entities or can be part of a larger group of plants (as with the hedge-bounded example). Monocultural agroecosystems, if not subdivided into plots, can be separated into planting units by age or management regime.

## Underlying Axioms

Generally, a more efficient plot is one that has the highest total outputs, reducing per unit (of output) harvest cost. A high degree of biomass will reduce weeding costs, more so if it involves biodiversity.

Also, gains can be expected in disease and insect control when multiple species cohabit in an area.

If two plants are highly complementary, then less space is wasted and the best results (LER) are obtained by growing them in close proximity. With total resource complementarity, the component species (two or more) should have a close interface distance and be evenly distributed. For a bioculture, where species *a* has a high degree of complementarity with species *b,* the best cross-section for an bicultural agroecosystem is ... *abababa.* .... Where complementarity is of the highest order, the best results are obtained using the optimal monocultural spacing for each species in the system. Again, these are interplanted and in close proximity. Less complementarity requires modification, involving a temporal shift and/or a spacing change.

A complete temporal shift, as opposed to a simultaneous planting, uses the species in full rotation. This is the strategy where transitory resources (light or water) are limiting, in-soil assets are abundant, and the intercrop does not depend upon immediate facilitatory effects (e.g., shading) for any productive gains. Rotations are also used where one species sanctions the buildup of an essential resource needed by a second species.

Semisequential planting (where the individual species temporally overlap for part of their growing periods) is a viable intermediate solution where the planting season and essential resources can accommodate the resource needs of an intercrop, but where some adjustment is needed. Where, due to short growing seasons, semisequential timing is not feasible, other alternatives, such as added nutrients or water, can adjust the resource use situation.

A decrease in planting density may be in order where resources cannot support higher yields and higher planting densities. In more resource-specific situations, where soil minerals are limiting, a spacing change may help in extracting the most from partial complementarity. Lower limits to density are imposed by a need for output levels and/or labor (harvest and weed control) efficiency.

Where a spacing change is justified can be illustrated with a bicultural example. Species *a* has the needed resources in abundance (e.g., a phosphorus-demanding species on a high-phosphorus site), while those resources needed for complementarity with species *b* are lacking (e.g., less abundant nitrogen for a nitrogen-demanding species). For the phosphorus-demanding species *a,* the best cross-sec-

tion spatial arrangement may be ... *aabaabaa* ... or, where excess nitrogen favors species *b,* so would an arrangement of ... *bbabbabb.* ...
Although this is documented for some prominent intercrops, such as maize-bean (e.g., Davis et al., 1987), this line of reasoning is not well researched.

Also possible are facilitative pairings, where in biocultural form, one species provides facilitative services to another, such that the system LER is above the 1.0 threshold. Common facilitative pairings include the use of cover crops and overhead shade trees. Common facilitative effects are improved soil nutrients, microclimate, nutrient and water retention, and other such services.

These guidelines continue to be refined. For most plant pairings, they are not easy to implement with the current level of understanding. Nonetheless, they do provide a sense of direction in agroecosystem design.

The landscape can be viewed as an expansion of individual productive units and resource-compatible complementarity. In many cases this may not extend, in any ecologically significant way, from individual ecosystems to the landscape. This extension does occur on a socioeconomic level for inputs such as labor distribution. When dealing with rotations, the larger landscape comes more into play, as a planting sequence must be assigned to productive units. Indirectly, this has greater ecological ramifications. Where the landscape is, in essence, one large biodiverse agroecosystem, complementarity has a greater role.

### *Landscape Complementarity*

The ability to thrive while growing in close proximity is the reason for many multispecies agroecosystems. Plant-on-plant bicultural complementarity, although seemingly narrow in scope, can be expanded throughout the larger landscape. Again, this can be spatial or temporal. The principles espoused also apply to two or three-plus polycultures.

Where species *a* is complementary with species *c* and species *b* with species *e,* some examples of resulting bicultures are ... *acacac* . .. or ... *bebebebe.* .... If species *e* lacks complementarity with species *a,* and species *c* does not grow well with species *b,* a viable landscape based on complementarity pairings can still be formulated.

A landscape can be formulated using distinct plots, thereby avoiding complementarity issues. In cross-section, this may be shown as follows:

$$\ldots acacacac \mid bebebebeb \mid acacacac \ldots \qquad (2.1)$$

where the plot boundaries ($\mid$) are used to separate noncompatible units. There are other ways to approach the same problem. In cross-section, for example:

$$\ldots abababababab\text{-}ebebebebeb\text{-}acacacacaca \ldots \qquad (2.2)$$

where the individual agroecosystems are connected (*b-e* and *b-a*) through species complementarity. An even more integrated example follows:

$$\ldots aabaacaabaacaabeebeebbacacababab. \ldots, \qquad (2.3)$$

which is less plot oriented, using agreeable pairings in a continuous, one-ecosystem landscape.

## PRIMARY AND SECONDARY SPECIES

Within plot-based agroecology, agroecosystems are composed of primary and secondary species. The output of the primary species is the most desired, whereas the secondary species may (1) only produce a low-cost alternate or subordinate output, and/or (2) have a facilitative role in promoting the growth or protection of the higher-value and more desirable species.

Complementarity is the key with any pairing. At the landscape level, the concepts of primary and secondary species are comparable, but with some differences. Landscape complementarity sidesteps a one-on-one ranking in favor of species importance. Most commonly, the high-value or staple crops, e.g., rice, maize, and wheat, head the list, with others following in order of relative importance.

A second landscape difference is that these plants can exist in separate areas, but still interact. For example, insect-repelling plants can be located adjacent to primary species. Although these may not con-

stitute a single agroecosystem and may lack plant-plant complementarity, the association may be beneficial.

## Multipurpose Plants

Multipurpose plants have only a single role in landscape. Commonly, these are primary species whose productivity is the one and
only purpose for their presence. As examples, wheat and maize may
provide only grain. They can transcend this by (1) providing a second
output such as wheat straw or maize stalks for forage, or (2) offering
facilitative services (e.g., as part of a rotational sequence that benefits
a succeeding species).

A basic premise is that, as component of an agroecological landscape, plants provide an output and/or facilitative service. Plants are
integrated into the landscape (in number and location) in relation to
the outputs and/or facilitative services provided.

### Productive Multiuse

Multipurpose plants may have a number of roles that are both productive and benefit the landscape. These are (1) specific-use and
(2) multiple-use plants. For example, the species *Uapaca kirkiana*
from southern Africa produces fruit and construction wood, has medical uses, and can be used as a cockroach repellent (Ngulube, 1995).
In addition to food and quality-of-life gains, the tree might have a
landscape purpose as a shade tree or to mitigate some environmental
problem. This is a particularly worthwhile multiple-use species that,
although few in number, figures prominently in the landscape.
Though this case is atypical, it demonstrates the possibilities.

Some plants have value and multiple uses, providing quality-of-
life gains such as medical benefits, spices, or firewood. Those with
household uses are often grown near where they are used. Others that
do not offer highly useful products (e.g., poles or forage) are less
prominently displayed. These plants may be greater in number, but
located in remote corners.

### Facilitative Multipurpose

Under this heading are plants that are primarily facilitative (with
some minor output) or purely facilitative (no usable output). A cover

crop located to mitigate erosion is an example of the latter. These plants can figure prominently in the landscape, but because they have few functional outputs and require little maintenance, often go unnoticed.

## DESIRABLE PLANT CHARACTERISTICS

The basic idea of desirable plant characteristics (DPCs) is that each plant has desirable agroecological properties, and plants are ranked, used, and placed based upon these. This is the foundation for a number of landscape concepts.

The concept of DPC advances the notion of the multipurpose plant, where a desirable characteristic is the ability to provide useful output or facilitative services. In addition, complementarity with a staple or high-value crop is a desirable characteristic. Usefulness is obvious in the case of primary (e.g., staple) crops, but DPCs have application more with finding a suitable secondary species.

### Characteristics

One use of DPCs is in pairing species to achieve complementarity. The range of niche relationships that are needed to achieve complementarity is not fully understood. However, there are some simple cases where complementarity exists on a physical or temporal scale.

Where a primary species is shallow rooted, a desirable characteristic, one that may provide complementarity, is a deep root system. Where a primary species requires more water and nutrients late in the growing season, complementarity may be assured by selecting a species that goes through a period of high growth early in the season.

Beyond complementarity, plants provide productive and facilitative gains. Productive output can be grain, fruit, wood, forage, etc. These subdivide into a host of categories. Grain and other food products have desirable attributes such as storage potential, taste, and ease of preparation. Wood and forage have their own desirable attributes.

Some of these attributes indirectly influence landscape use. For example, a variety with good-tasting fruit may be less productive and/or require better soils than another variety with less palatable fruit. Other outputs may have a direct influence. Forage may be ranked, from highly palatable to inedible, for each type of animal (e.g., horses, cows, goats, pigs, etc.). Pasture management can be enhanced and potential damage mitigated based on the DPCs of the constituent species (trees, shrubs, grasses, etc.). This aspect of DPCs sways both the component agroecosystems and/or the overall landscape design.

## *Ranking of Characteristics*

The list of desirable plant characteristics is long and detailed. Detailed refers to cases where one DPC can be subdivided, e.g., palatable forage can be further ranked into the nutrient content, effect on animal digestion, and other fodder properties. Some of the bases for these rankings are listed in Table 2.1. A formal mathematical method for ranking is discussed by Wojtkowski (1998, p. 97).

## *BIODIVERSITY AND AGROBIODIVERSITY*

The advantages of biodiversity are well entrenched in ecological thought. However, there are some differences in this concept as applied to agroecology. Agrobiodiversity (or agrodiversity) centers on vegetation with direct application to productive processes, including facilitative species. Biodiversity places more emphasis on those plants that provide environmental services and contribute to formulating an effective ecological landscape.

Agrobiodiversity is concerned with the diversity associated with common crops. Most crops have a vast number of varieties and closely related species. These may appear similar in common use situations, but may contain sufficient, if unrecognized, genetic advantage in countering risk or when intercropped. Seemingly small differences can be reflected, and find favor, through one crucial DPC and/or use in matching a plant species with a growing site.

Agrobiodiversity also applies to myriad species with useful DPCs. There is no clear line between natural diversity and agrodiversity; agrodiversity can be locational—a weed in one place, a facilitative plant in another. The point is that an agroecological landscape cannot always be considered a storehouse for maintaining natural biodiversity, much as a natural landscape will not automatically conserve

TABLE 2.1. Desirable (common) characteristics for companion species

| General attributes | Specific characteristics | | |
|---|---|---|---|
| | Shade trees | Superbiomass plants | Cover crops |
| Resource compatibility with the primary crop | Valuable or useful secondary outputs | High aboveground biomass production | Does not climb on taller plants |
| An ability to grow on poor soils | Smooth bark that does not harbor herbivore insects | Moderate to high, balanced foliage nutrient content | Short statured |
| Tolerates climatic variation | Self-pruning with good bole form | Good burning properties (depends on use) | Produces ground-level, heavy shade (through high biomass or large leaves) |
| Ease of establishment | Limited maximum size for tree growth | A dense network of fine roots | Provides forage or an alternate product (e.g., bean) |
| Freedom from pests and diseases | Strong branches and stems (blowdown resistance) | Absence of toxic substances in the foliage | Drought and frost resistant |
| Lacks the capacity to become a weed | Small leaves to minimize raindrop coalescence and drip damage | High leaf-to-stem ratio | Fits within the temporal sequence of the primary crop |
| Lack of root-suckering properties | For deciduous trees, rapid flushing of the leaves to regenerate shade | Small leaves or leaflets for rapid decay or larger leaves for better erosion or weed control | Has allelopathic properties to prevent weed seed germination |
| Ability to trap nutrients (nutrient net) | Small canopy to reduce wind resistance and ease tree harvest with minimum crop damage | Leaves that detach readily | Promotes a microclimate to speed the decay of residual vegetation |
| Ease of control and eventual elimination | Open canopy to allow greater light penetration | Leaves that are highly, moderately, or not palatable | |
| Spinelessness (spines can also be a desirable property) | High salvage value (high wood or firewood value) | An appreciable nutrient content in the root system | |
| A high rate of nitrogen fixation | | | |
| Dry-season leaf retention (tropical plants only) | | | |
| A preponderance of deep roots | | | |

*Source:* Compiled by Wojtkowski, 2002, p. 110. Shade trees: modified from Beer, 1987. Superior biomass plants: Young, 1989b; Rachie, 1983; Beer, 1987.

agrobiodiversity. If natural diversity is to be encouraged in an agro-diversity forum, native plants are best promoted through DPCs.


## DESIRABLE AGROECOSYSTEM PROPERTIES

As with individual plants, ecosystems also have definable properties. In an agroecological landscape, production is not the only goal. Agroecosystems should have a range of desirable properties, some contributing to the economic outcome (e.g., lower costs), others adding an environmental return.

The concept of desirable agroecosystem properties (DAPs) influences the placement of various agroecosystems with regard to their known attributes. They can be absolute, as with site attributes, or relative, in relation to the properties of neighboring systems.

Outside of the productive considerations, DAPs may include the ability to anchor soils, bioremediation aspects (where a site is improved through agroecosystem placement), frost resistance, drought resistance, etc. This topic is important in landscape agroecology. Some DAPs are listed in Table 3.1 and further developed through specific use discussions (Chapters 6 through 8).

### DAPs in Agroecosystem Design

DAPs are a planned outcome that succeeds through knowledge, study, and formulation. This can involve trial and error to arrive at an agroecosystem that has the desired properties. As field of study, this approach is new and, although only a few articles and texts have directly addressed this topic (e.g., Wojtkowski, 1998), numerous sources advance specific ideas, concepts, and components.

As biodiversity grows, a system becomes more than just the sum of its parts, and the list of DAPs grows. Less biodiverse systems still have internal dynamics and DAPs, and these can be expanded through design. The idea is to use the ecological tools nature provides to realize DAPs. If this is not possible, then a more conventional practice is employed. For example, weeds may be suppressed by ecological means (i.e., allelopathy, shade, etc.), but if the plot design does not support this, hand weeding may have to suffice.

To achieve the full range of desirable properties, some compromises are in store. The process usually starts with a primary species and continues with three approaches that provide direction:

1. Companion species (one or more) with the needed DPCs
2. Spatial patterns (both horizontal ground level and vertical canopy)
3. A sustaining temporal sequence

One or all may be employed. As the basic building block, plant/plant complementarity confers a range of desirable properties, although other more competitive approaches can also be used.

This is not a well-understood undertaking, often necessitating trial and error in development. The agrotechnologies (see Chapters 3, 4, and 5) reduce trial and error by providing starting points in an otherwise murky process.

## LAND USE INTENSITY

Clearly, land use intensity varies between regions. At one end of this scale are natural ecosystems that are barely touched by humans in their hunting and gathering activities. These regions usually have very low population densities. At the other end of this spectrum are sectors where population density is high, agriculture is intense, and all available plots are under cultivation, some compelled to provide multiple yearly harvests.

What occurs between these extremes can be difficult to quantify (Shrair, 2000). This may be, in part, because different alternatives exist.

An intermediate point is where a portion of the land, usually the best sites, is intensively cultivated while the remainder is used, but with far less rigor. This may occur because of climate (i.e., rocky and/or dry hillsides with fertile, well-watered valleys), farm machinery (i.e., costly tractors that can only be used in limited areas), dietary needs (e.g., staple crops that only grow in certain locales as with paddy rice), or for other reasons.

In the opposite case, the land receives equal treatment, i.e., the per plot inputs are more or less equal. Again, this can originate with climate and topography, types of agriculture or forestry practices, and/ or other cropping needs (e.g., irrigation limits).

## INTERAGROECOSYSTEM EFFECTS

A farm or forestry landscape can be a series of independent plots or agroecosystems, each formulated to be self-contained without a need for interplot interactions. The opposite situation is where ecological dynamics are acquired from other (agro)ecosystems.

This can be through the specific interaction zone (SIZ). The SIZ is where the ecological dynamics (often a DAP) that occur in one (agro)ecosystem overlap with neighboring systems. The overlap can be slight or of considerable distance and/or intensity.

There are some broad variations on the theme of the SIZ and agroecologically interdependent landscapes. The variations involve how an ecological influence is transmitted across wide areas. Two methods are (1) augmentation and (2) expansion. Using wind effects, these are illustrated in Figure 2.1.

### Ecological Augmentation

Ecosystem ecological augmentation is used when, for a productive agroecosystem, some important DAPs are lacking or weak and need to be reinforced. Neighboring ecosystems, enhanced in the ecological property needed and with an SIZ that extends outside that ecological system, augment the weak or lacking property. That is, any internal DAP attribute (e.g., insect control, wind resistance, etc.) can be strengthened through the manipulation of external ecological forces.

One of the key concepts in landscape agroecology is that plot size can be used to manipulate the agroecological forces operating within the larger economic entity. If plots are large, the amount of interplot dynamics is small. This is especially true if the SIZ for the DAP in question does not extend far.

If plots are small and/or shaped such that there is a large amount of external plot surface area, then the external forces generated by neighboring systems have a large impact on the internal plot dynamics. For example, if a plot has poor frost resistance, it can be increased by the design of surrounding plots (i.e., dense, high canopies).

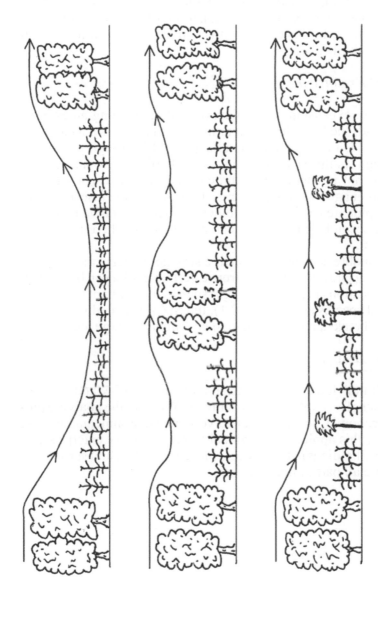

FIGURE 2.1. Three wind control scenarios. The upper has large areas where the limited windbreak SIZ (see text) does not protect the entire area. The center d'awing shows augmentation, where an added windbreak augments those already present. The lower drawing demonstrates expansion, where an external SIZ is expanded through added in-field vegetation. (Note that these are not drawn to scale relative to windbreak SIZs.)

Figure 2.1 (middle) shows augmentation in a wind control setting where an added windbreak, intermediate in a field, augments those on the periphery. In this case, subdivision has increased the area of the windbreak SIZ.

## *Ecological Expansion*

Ecological expansion, rather than only strengthening an external DAP, expands the DAP through internal redesign across a wider area. The purpose is to increase the overlapping SIZ through productive unit redesign. In this case, a strong external effect (from a neighboring plot) is not intensified but, through redesign, is carried further into the target agroecosystem. For example, favorable insect dynamics in one plot are carried into another through vegetative modification in the second. This strategy can prove useful where large plots are the norm.

In the windbreak example, there is a stronger effect at the edge of a plot. Within a large area, the SIZ from a perimeter plot can be lost but, with modifications in internal design (such as intermittent hedgerows or parkland species as in Figure 2.1), the SIZ associated with an external windbreak can be expanded to encompass a large area.

## *Landscape Design*

As with individual ecosystems, a landscape can be, and often is, the recipient of an explicit design strategy. Whether the goals are productive, economic, social, and/or serve some environmental purpose (or whether this applies to individual farm or forestry enterprises or crosses land holdings), the concepts presented in this and subsequent chapters are all part of this process.

# Chapter 3

# The Agrotechnologies

The one-plot one-agroecosystem model is a widely observed landscape type, where the agroecosystem is also a recognized agrotechnological variant. Within the one-plot one-agrotechnology model, the agrotechnologies are the key building blocks in constructing an agrotechnology-based landscape. This view is widely subscribed and forms the basis of understanding that underlies a large segment of research and extension.

## BASIC CONCEPTS

An agrotechnology (agricultural technology) is a land use practice that addresses a distinct productive, environmental, or socioeconomic need and/or overcomes a site limitation. An environmental problem can be some form of erosion; a common socioeconomic need is the production of staple crops; and a site limitation may involve improving the quality of nutrient-poor soils.

The agrotechnology achieves distinctiveness through unique characteristics. A unique attribute can be spatial, temporal, utilize distinct ecological dynamics, or be described through some other ecological quality. The component species (one or more) and complementarity between plants does not define an agrotechnology. It is defined only by the land use problem dealt with.

The use of spatial patterns to address the problem at hand is a common rationale to define an agrotechnology. Although most agrotechnologies are described at a moment in time, there are temporal considerations and temporal agrotechnologies, where the transitory state can also address distinct productive, environmental, or socioeconomic needs or deal with a site limitation. The temporal sphere can therefore also define an agrotechnology.

Some unique agrotechnologies have evolved in response to unusual situations. For example, the Aztecs of early Mexico had floating gardens that allowed for the production of terrestrial crops on shallow lakes. The defining attribute was soil-covered floating mats. Although interesting for purposes of illustration, this agrotechnology has limited application.

Most examples are not as dramatic. An intercrop can be differentiated from a monoculture through the unique ecological dynamics contained. Further, intercrop plantings and harvests are more involved, and the resulting ecological complexity offers unique applications (e.g., insect and weed control) that monocultures cannot fully address through internal processes alone.

An agrotechnology encompasses a primary species, spatial or temporal patterns, and management inputs (i.e., types of plowing, pruning, application of inputs, etc.). The primary species generally provides the most valuable or useful outputs, although in highly complex systems (e.g., agroforests) a number of species can share the title of primary species.

An agrotechnology may also have an environmental function and, if this role is important enough, it may transcend any immediate productivity concerns. In some cases, there may be outputs, but no clear primary species. This often exists with naturally occurring outputs in riparian buffers or other wild areas.

A complete discussion of internal dynamics of the different agrotechnologies is outside the parameters of landscape ecology. As components, each agrotechnology contributes within the ecological landscape.

## NONTEMPORAL PATTERNS

The spatial pattern often describes how static or nontemporal agrotechnologies (those existing in a moment in time) are put into practice. The six basic patterns frequently mentioned (see Figure 3.1):

1. Blocks
2. Borders
3. Strips
4. Groups (or clumps)

5. Individual plants
6. Rows

It should be noted that these are generally associated with bicultures and are not the only view of spatial patterns.

Wojtkowski (2002, p. 52) has divided patterns into fine and coarse where use is based on plant-plant complementarity. Briefly stated, the fine patterns (upper row, Figure 3.1) have more interspecies interface and are used to take advantage of a high degree of complementarity between component species. The coarse patterns (lower row, Figure 3.1) are used when less complementarity is evident. For agroecosystems of three or more species (three-plus polycultures), these patterns are less appropriate, but may still guide the design process.

Beyond these patterns, there are pattern arrangements, including the actual dimensions, the planting densities for the various species, and other installation imperatives. For example, when initiating a row system, the rows can be single (where the component species alter-

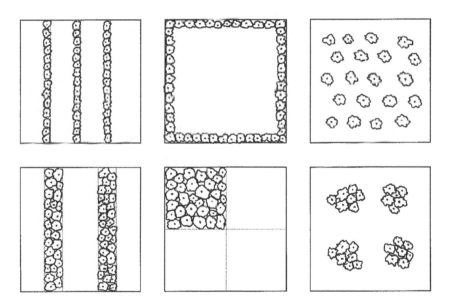

FIGURE 3.1. The six basic spatial patterns illustrated through the location of the secondary species. These are, from left to right, top row, (1) row, (2) boundary, and (3) individual; bottom row, (4) strip, (5) block, and (6) cluster. (*Source:* Modified from Young, 1989a.)

nate rows) or double (two rows of one species, then two rows of another), vary as to internal planting density, vary as to the number of rows per area (which sets the interspecies interface distance), and/or have rows oriented in a particular direction. These do not substantially alter the nature of an agrotechnology and are not defining attributes.

Agrotechnologies with more than one species are also multidimensional, involving the height relationship between component species. For example, hedgerow and tree row alley cropping systems employ the same spatial pattern (a row pattern) with height differences (see Photo 3.1). The resulting internal dynamics are enough to change the nature and purpose of the system and partition them into separate agrotechnological categories.

## TEMPORAL PATTERNS

Most of the agrotechnologies emphasize static designs, those with spatial pattern and presence. Agrotechnologies can also be defined separately in the temporal sphere.

This facet should not be overlooked. The temporal agrotechnologies offer an alternative to landscape design where, because of land use limitations (holding size and/or topography) and cropping needs, spatial patterns alone cannot fully address agroecological need. The rotation patterns that define the temporal sphere are discussed in Chapter 5.

If a landscape is designed using both the temporal and spatial planes, more alternatives are provided and more can be accomplished. The tradeoff is in the added ecological complexity that accrues.

## DESIRABLE CHARACTERISTICS

Each agrotechnology (spatial or temporal) has a set of desirable agroecosystem properties (DAPs). Some of these are an offshoot of the design package (discussed later in this chapter). Other DAPs involve the role of the agrotechnology within the overall landscape and an agrotechnological relationship with neighboring systems.

PHOTO 3.1. An alley crop system with maize between rows of natural vegetation. This is also a transitory phase that, once the newly planted tree stems grow, evolves into another agrotechnology (e.g., row cropping and heavy shade systems). The tree block in the background is representative of the tree plantation that will end the sequence.

## Ecological Properties

Agrotechnologies can be classified by their relative merit in a number of categories. This concept was first presented in Chapter 2 and is expanded here. A list of basic DAPs is given in Table 3.1.

Ecosystems are subject to any number of natural stresses (as listed in Table 3.1). Some events are more common than others, but any meaningful countermeasures depend on selecting an agrotechnology that can resist some level of stress and/or adding a positive character-istic (either internal or external) that can reverse an imposed negative influence. Some of these are briefly discussed here.

Drought resistance varies, but outside of species moisture needs, agrotechnologies can be roughly ordered by their relative resistance to a shortfall in precipitation. Where some retain moisture well, oth-ers loose it quickly. The opposite of drought is a rainfall deluge. A system can remain unchanged by a short period of high rainfall or, if it is defenseless against this stress, crop yield (and soil) can be af-fected. Again, some agrotechnologies are formulated specifically for, or have an innate ability to confront, this problem.

Wind can cause a host of negative (and a few positive) effects (as discussed in detail in Chapter 7). A number of measures confer wind resistance in agroecosystems.

TABLE 3.1. Desirable agroecosystem properties against natural stresses

| Natural landscape stresses | Rotational (postharvest) site attributes | Economic outcomes |
|---|---|---|
| Drought | Insect balance (type and population levels) | Land equivalent ratio |
| Rainfall deluge (including lowland flooding) | In-soil diseases | Relative value total |
| Winds | Weed infestation (in-soil potential) | Cost equivalent ratio |
| Detrimental insects (in-cluding those that assail farm animals) | Crop residue | Economic orientation ratio |
| Diseases | Soil nutrients | Time value |
| Fire (uncontrolled) | Soil physical properties | Risk |
| Weeds | | |
| Temperature severity (heat or frost) | | |
| Soil extremes (plant in-hospitable factors) | | |

Hungry herbivores can find crops to their liking, or the environment can be unpleasant with lurking predators and/or repellent plants. Predators can be an external or internal influence, while repellent plants are generally an internal factor. Outside of these, other design alternatives can control aboveground and belowground insect populations.

Similarly, diseases can be encouraged or controlled within an agrotechnology. This may, in part, be regulated by microclimate, where a shady environment may be more conducive to disease spread than a less humid, more open situation. Timing and rotations are also a control tool.

Severe fire can be destructive to crops or trees, or fire can be used as a postharvest tool against insects, diseases, and weeds. Unwanted plants are costly to remove, and ecosystems can be formulated (through biodiversity, biomass, or other measures) to discourage weeds.

An agrotechnology and its component plants may have the ability to counter undesirable soil characteristics. This includes overcoming aluminum toxicity, salt problems, and soil acidity or alkalinity. This is often done through resistant species matched in an appropriate agrotechnology. For most nutrient-deficient soils, rotations, rather than specialized designs, can adjust the in-soil nutritional balance for subsequent cropping phases.

*In-Place Stress Control*

How well in-place agroecosystems handle natural stress can be a measure of productivity and risk in a particular cropping area. Having a mix of areas with highly diverse DAPs reduces risk. Having interspersed areas where the SIZ for risk factors extends outside the individual areas also diminishes risk.

The first criterion for natural stress management is to locate agroecosystems that are resilient to or can counter an effect whenever and wherever the influence is greatest. This is obvious. For example, on steep slopes, it is better to utilize agroecosystems that have great erosion and water management properties.

The second criterion, interspersed ecosystems, is to form a pattern mosaic (spatial and temporal) where any migrating or transitory risk that can affect susceptible agroecosystems is halted at a nonsusceptible ecosystem and not carried across the full landscape. For example, insect movement is hindered (1) spatially by the placement of non-palatable crops and (2) temporally through crop rotations designed to interfere with the reproductive cycle of herbivore insects.

A third criterion, the SIZ, is valid where a stress influence extends outside an ecosystem plot. A common example is to use a tall, wind-resistant ecosystem as an upwind neighbor to a wind-susceptible crop.

These criteria underlie much of landscape design and, if individual ecosystems are arranged properly, the risk reductions can be achieved without much additional effort. Where plot arrangement does not take into account any interplot potential where risk is magnified (e.g., with a farm having only staple crops with no supporting vegetation), additional protective structures may have to be added.

Without additional structures, an ecosystem-diverse landscape can be mutually reinforcing. The SIZ occurs when an agrotechnology is susceptible to a natural stress. A neighboring plot, one which counters this stress, can positively influence the plot in question through ecological augmentation.

## *Rotational Attributes*

Any system, after completion (harvest), leaves a site with certain properties. Ideally, the site will be erosion proof, have superior soil characteristics, and be free of herbivore insects. After harvest, a site may not always be perfect for the next rotation, but some positive attributes are expected, e.g., with in-soil organic residue that can help hold moisture or soil devoid of in-soil insect eggs or larvae. Any subsequent cropping or interperiod treatment (e.g., fire) is decided based upon that which remains.

Equally important is the resulting nutrient balance. Where imported nutrients are not an option, this balance determines what crop type will best follow. For commercial activities where fertilizers are used, this determines the best mix of elemental nutrients to apply. Included in this end stage calculation is soil pH, bulk density, and other such attributes. These are used in rotational decisions and can be classified as an ecological property.

## **Economic Properties**

The economic landscape characteristics of individual agroecosystems are best defined through qualitative measures (e.g., profitability). These are as important, or even more so, in determining system use than the ecological or stress control characteristics. These properties are listed in Table 3.1.

Monetary criteria work well with simpler systems, but can be lacking when applied to highly biodiverse, multiple-output systems (e.g., agroforests) which, to date, have defied comprehensive analysis. This is due to the inclusion of nonmarketable and unvalued outputs, the infrequency of production from some species, and the use of casual, unplanned, and difficult-to-quantify inputs. An incomplete economic picture can be a barrier to use in commercial land use enterprises unless based upon firsthand experience.

With any agroecosystem comparison, one method may not suffice. Different economic measurements provide a fuller picture of the effects and gains from different land use alternatives.

## Land Equivalent Ratio

The LER is the most basic ecological ranking and determines how efficiently a cropping combination uses available on-site essential resources. As a measure of complementarity, this is both an ecological and economic measure. The LER for a two-species system is calculated as

$$LER = (Y_{ab}/Y_a) + (Y_{ba}/Y_b) \qquad (3.1)$$

where $Y_a$ and $Y_b$ are the monocultural yields of species $a$ and $b$, respectively. $Y_{ab}$ is the output of species $a$ grown with species $b$, and $Y_{ba}$ is the output of species $b$ in combination with species $a$.

For comparison, a ratio greater than 1 is ecologically superior to the monocultural controls ($Y_a$ and $Y_b$) and indicates some degree of plant-plant complementarity. A comparison ratio of less than 1 indicates a competitive situation where the monoculture may be the better option. With this and other measures, these are intertwined with spatial theory (e.g., Wojtkowski, 2002, p. 50) and are a major consideration in agroecological development.

## Relative Value Total

The relative value total (RVT), with an added monetary element, is more economic and less ecologically inclined than the LER. The equation is

$$RVT = (p_a Y_{ab} + p_b Y_{ba})/(p_a Y_a) \tag{3.2}$$

where $p_a$ and $p_b$ are the output or market prices for species $a$ and $b$, respectively. The monoculture with the most revenue potential is the denominator, i.e., $p_a Y_a > p_b Y_b$. Usually $Y_a$ is the primary species.

## Cost Equivalent Ratio

Another economic measure is how well management inputs are used. These are evaluated in financial terms through the cost equivalent ratio (CER).

$$CER = (C_a/C_{ab})(RVT) \tag{3.3}$$

With this equation, $C_a$ is the costs associated with a monoculture of the primary species $a$, $C_{ab}$ is the costs connected with the polyculture (species $a$ with $b$) where a value greater than 1 indicates the two-species polyculture is more efficient with management inputs (costs). A value of less than 1 occurs when the monoculture is the more cost-efficient option.

## Economic Orientation Ratio

In combination, the RVT and CER are useful on the landscape level. This is economic orientation. An agrotechnology may sacrifice some potential yields (income) for a correspondingly larger reduction in inputs (costs). These are cost-oriented systems. Conversely, more inputs can be added to achieve greater outputs. These are the revenue-oriented systems. The combined RVT and CER is the economic orientation ratio (EOR). This is expressed as

$$EOR = RVT - (C_a/C_{ab}) \tag{3.4}$$

In contrast to other economic measures, EOR does not indicate productive or economic superiority. Instead, the EOR shows how the economic gains or losses of a polyculture are achieved when compared to a monoculture of the primary crop or other design variation. This has strong landscape implications and is discussed in Chapter 10.

*Time Value*

In any productive enterprise, the temporal interval for the outputs is important. Land users want more immediate and/or continued outputs. Commonly, land users will reject agroecosystems with high, but far future, outputs or those with superior ecological properties in favor of systems with lower, but more immediate, returns.

Simple, nontemporal comparisons are made using the cost of inputs subtracted from the market value of goods produced (i.e., revenue – costs = returns or profits). There are two ways of determining future value: net present value (NPV) and the least common denominator (LCD).

*NPV.* Time value is a monetary determination estimated through the use of NPV. The NPV employs a discount rate *(i)* to determine the present value of future income. The calculation is

$$NPV = -C_0 + (R_1 - C_1)/(1 + i) + (R_2 - C_2)/(1 + i)^2 + (R_3 - C_3)/(1 + i)^3 + ... + (R_n - C_n)/(1 + i)^n \qquad (3.5)$$

where R and C are revenue and costs respectively for the years 0 to *n*. This is a very common business evaluation technique and is found as a subroutine in all business computer packages. To provide a balanced picture, the calculation and comparison should be made using both a standard and an undiscounted (i.e., 0 percent) rate and a decision made accordingly. NPV can also be used with the LER, CER, and other indices, again to equate time periods.

Besides comparing income from different time periods, this calculation is also helpful to compare uneven cash flows from different systems. For example, one ecosystem has an initial $200 per area planting cost, a $2,000 maintenance cost in year 10, and harvest value in year 30 of $10,000. In contrast, a second system has no initial planting cost, a maintenance cost of $4,000 in year 10, a small cash inflow of $2,000 in year 20, and a harvest value of $20,000 in year 50.

The financially superior system may not be immediately intuitive. For the first system, the NPV at 4 percent (the *i* value) is $1,532.05 and, for the second, the 4 percent NPV is $1,024.77.

Although of use, there is some danger, as a high discount rate can induce a bias on planting highly beneficial systems that take considerable time to mature. For these examples, the NPV with an *i* value above 8 percent is negative.

It may be advisable to study the situation using a number of criteria before making a final decision. In this example without NPV (or with a discount rate of 0 percent), the first system has a value of $7,800 and, for the second, the net value is $18,000. In this example, the 0 percent rate reverses the recommendation using a 4 percent rate. This may not always be the case, but this points out some of the dilemma using a single criterion. Beyond time value, numerous other factors can come into play. These can supersede NPV analysis, clarifying or clouding an issue.

*LCD.* The second method, the LCD, sums yields for the different time periods until a common time period unit is reached. For example, three rotations for a two-year cropping cycle and two for a three-year cycle have a common time period of six years. The system (the two- or three-year cycle) that gives the best yields or income summed for this six-year period is judged the better.

Both LCD and NPV do not take into account ecological and other intangible benefits. An example is sustainability, which has a strong temporal component (Hansen, 1996; Rigby and Cáceres, 2001). For this, the NPV at a high discount rate (with the emphasis on earlier rotations) may give misleading results. To overcome this, the more distant time periods for the extended rotations may be directly analyzed to assess sustainability without considering the earlier rotations or LCD.

*Risk*

Another economic component is risk. Natural landscape stresses (drought, floods, insect and disease outbreaks, high wind, etc.) vary greatly. The conventional economic assessments of risk seem to miss the mark as there are, in addition to the different forms of risk, different perspectives on their relative importance and many unknowns on how agroecosystems cope with unexpected variation. As such, evaluation is mostly a land user affair.

For any natural stress, factors include frequency (unusual or reoccurring), timing (how long), severity, the ability of the agroecosystem and landscape to withstand the stress, the importance of the natural stress to the land user, and the perception of the potential for damage. These can also cloud any quantitative determination.

The agroecosystem is protected through a number of mechanisms, which are detailed in subsequent chapters. Most often, the protection is not total, only reducing risk to an acceptable level. When additional protection is introduced from outside a particular system, the protection can be more inclusive.

## AGROTECHNOLOGICAL CLASSIFICATION

The purpose of an agrotechnology is to partition the ecological knowledge base into manageable segments and, through this, promote wider use. Basically, it is easier to research a system formulated for a specific purpose, containing clearly specified plant roles and ecological relationships. With precise DAPs, the process of fostering wider use is expedited.

### Methods

Within the framework of the agrotechnology, subclassifications abound (e.g., Sinclair, 1999; Nair, 1990). Already mentioned is the use of species content; e.g., a wheat monoculture differs from a maize monoculture. Agrotechnologies may also be ordered by the internal spatial pattern (e.g., row, strip, clump, etc.), temporal content (annual, perennials), short- or long-term ecological properties (e.g., the capability for accommodating natural flora and fauna, enriching soil, etc.), ability to counter natural stress (as mentioned earlier in this chapter), economic properties, and/or other criteria.

### The Design Package

The DAPs may provide a starting point in landscape design where the focus is on ecological characteristics. Land users require more than agroecological compatibility between plots and can delve deeper into those attributes that match an agrotechnology with use. Three components form the basis for a design package:

1. The land use problem addressed
2. The site requirements, including soils, climate, rainfall, topography, etc.
3. The socioeconomic situation, which includes the level of farm intensity, labor availability for the proposed system, land use requirements, etc.

A design package is formulated by tallying the different classification methods. For example, the desired system may focus on productivity, be revenue oriented, be part of agroforestry, have yearly rotations and minor facilitative effects, be suitable for a slight grade, and be directed toward the production of one staple crop. A number of agrotechnologies fit this description, including hedgerow or tree row alley systems. The output mix and tree crop compatibility further refine the choice.

Through selection of an agrotechnology and the corresponding choice of a component species (one or more), an agrotechnology conforms to a specific need within the larger landscape. For this to be successful, the three elements mentioned and their subdivisions must agree. That is, if one element is needed (e.g., the ability to improve the soil), the agrotechnology must support this element.

Another set of criteria, one that encompasses the type of landscape desired, involves where the ecological emphasis lies. For this purpose, the agrotechnologies can be divided into three broad categories. The classification used in this text is based on whether a system is

1. principal-mode,
2. temporal, or
3. auxiliary.

A fourth category is derived by combining, within each category, different principal-mode or temporal agrotechnologies. For example, hybrid agrotechnologies can be derived by combining principal-mode systems or by combining temporal agrotechnologies. By definition, hybrid agrotechnologies are not created by mixing temporally static principle-mode and transitory temporal technologies.

Principal-mode and temporal systems are not exclusive; the transient phases of a temporal agrotechnology can be different static principal-mode systems. Temporal phases are another agroecological tool used to achieve the desired outcomes.

## Principal-Mode Agrotechnologies

Principal-mode agrotechnologies are responsible for most of the productive output of a land use enterprise and constitute the principle means of production in forestry and agriculture. The distinctiveness of each lies in the mix of DAPs and the design package each offers.

For any system, there is an establishment period. Typically, these systems are described in their static and mature form and are best viewed as points along continua rather than discrete, nonconnected systems. The presently documented examples, in essential form, are listed here and described in the next chapter. As more information is gained, this list continues to grow.

Absorption zones
Agroforests
Aqua-agriculture
Aquaforestry
Alley cropping (hedgerow)
Alley cropping (tree row)
Entomo-systems (insect)
Forage (feed) systems
Intercropping (multiseasonal)
Intercropping (seasonal)
Isolated tree
Microcatchments
Monoculture (perennial)
Monoculture (seasonal)
Parkland
Root support systems
Shade (heavy)
Shade (light)
Strip cropping
Support (perennial)
Support (temporary)
Terraces (constructed)
Terraces (progressive)

These systems are differentiated by purpose. For example, moderate and heavy shade use the absence of light for distinct purposes, and are divided accordingly. In contrast, live fencing has discrete forms that serve the same purpose and are variations of the same agro-technology.

## Temporal Agrotechnologies

The second group are the temporal agrotechnologies. In simple form, these may be a single-season crop or a repeating series of seasonal crops. In a more complex arrangement, these can be a logical progression where three or more principal-mode agrotechnologies are ecologically and economically connected. The temporal agro-technologies are listed here (with some common subdivisions) and described in Chapter 5.

> Sole cropping
> Rotational cropping
> Fallows
>> Facilitative
>> Productive
> Overlapping patterns
> Taungyas
>> Simple
>> Extended
>> Multistage
>> Final stage

## Auxiliary Agrotechnologies

The third group is the auxiliary agrotechnologies. They lack any or have only a minor productive role. The chief reason for their existence is ecological facilitation with or between principal-mode agro-technologies or to serve some other environmental purpose within the landscape (e.g., clean water). They are detailed in Chapter 5. Natural ecosystems, such as forest fragments in a managed landscape, can be classified as an auxiliary system. Because they are not human fabricated nor human managed, they are not included here.

> Biomass banks
> *Cajetes*
> Catchments
> Infiltration barriers

Firebreaks
Living fences
Riparian defenses
Water channels
Waterbreaks
Windbreaks

Many principal-mode agrotechnologies accomplish similar eco-logical tasks, while serving a productive role. Finding a principal-mode system with the needed DAPs is not always possible, so, to fill the ecological need, an auxiliary agrotechnology is employed.

In theory, auxiliary systems can have a temporal sequence. In prac-tice, this may weaken their design intent and may run counter to the DAP for which one is selected. An exception may occur when an aux-iliary system is linked with a temporal agrotechnology, and the evolu-tion of the auxiliary system (and the unfolding DAPs) parallels that of the developing temporal system.

*Other Divisions*

In addition to the three categories described, some interesting ad-ditions and options are obtainable through a redesign of the basic agrotechnologies These can and do find use.

*Doubling.* An agrotechnology can be formulated as a double-pur-pose system, combining some but not all design features of the two. This can be done without changing the intent or compromising the design purpose of either component system. These systems generally do not cross categories (i.e., principal-mode, temporal, or auxiliary) with the proviso that location (topography, soils, etc.) and use (for crop production or as supporting systems) suit both.

An example is a riparian (auxiliary) that can be a biomass bank (also auxiliary) or a parkland (principal-mode) that can have absorp-tion zones associated with the trees (principal-mode). Given location and use constraints, the opportunity to find doubling opportunities is not overwhelming, but still adds to the array of available land use op-tions.

*Hybrid.* The use of hybrid agricultural technologies is another landscape option. Hybrid agrotechnologies are a combination of two separate, usually principal-mode systems. The purpose in doing so is to augment the DAPs of one with those of a second.

For example, a hedgerow alley cropping system can be combined with a parkland system. Although hedgerow alley cropping will gain little in water-soil erosion prevention, some gains are possible in insect and wind control and in overall productive capacity. This may require some modification in the design of one or both to accommodate the expanded objectives. For example, the parkland tree can be part of, or disconnected from, the hedgerow.

Many combinations have hybrid potential depending upon the amount of use compromise that is acceptable. Given the number of existing agrotechnologies and the lack of in-field examples, the use of hybridization may seem more an abstraction than an alternative. Despite this, it may be an option to be considered.

*Supplementary Additions.* For many agrotechnologies, it is possible to add useful flora or fauna without substantially interfering with the primary purpose. Rattan vines in the canopies of rubber trees and bamboo, if well managed, add value without intruding upon the productive purpose (Modh Ali and Raja Barizan, 2001). Other examples are truffles below tree plantations and edible insects in perennial systems. The reason for these additions is to add value without compromising productive or ecological purposes.

## DEVELOPMENTAL STAGES

In the development of a new or revised agrotechnology, a number of stages may occur as a result of field use or as a product of a research station. For many, the origins are lost in time; others can be traced to a specific idea or need. An example of the latter is hedgerow alley cropping, which started by planting young nitrogen-fixing tree species in maize fields and, through time, evolved into its present form (IITA, 1976-1982).

The ending stage is an technology well formulated for use along with a number of notable variations. Identifiable steps along this road are the (1) idea or draft, (2) agrotech, (3) refinement, and (4) expansion.

## *Idea or Draft*

The process starts when a land use need is identified and some ideas are proposed to address it. From a formal, more scientific perspective, this involves a critical look at the options, taking into account novel ecological concepts and untried methods. Economic analysis can determine minimum acceptable productive and cost figures. The balance (inputs versus harvests) in essential resources can be studied to see if the proposed idea is sustainable. If not, then variations are looked at (of species, rotations, planting densities, etc.) to see if the initial idea can be improved upon.

## *Agrotech*

At this point, a prototype and recognizable agrotechnology or agrotechnological variation has emerged. Some of the criteria for successful use exist; e.g., some primary and secondary species have been identified, the spatial and temporal patterns are known, and there is demonstrated potential. Considerable development remains, but with the confidence that this will prevail on farms and in limited adoption.

## *Refinement*

At this stage, the agrotechnology is performing as desired, but more fine tuning, especially on management variables (e.g., planting and pruning times), is required. Land users are learning how to manage it and have begun to actively modify the system. The different participants may be experimenting with other secondary species and/or are improving upon the initial management inputs.

## *Expansion*

As the success is documented or other land users see the results, this initiates the expansion phase where use parameters are examined, different component species are tested, and the potential for use in other socioeconomic situations, climatic zones, and sites is explored. In this phase, notable variations are observed and documented.

# Chapter 4

# Principal-Mode Agrotechnologies

Principal-mode agrotechnologies are those that are formulated to produce one or more outputs. For definitional purposes, they exist at a moment in time without temporal conditions and are usually described in their developed, fully functioning form.

Landscape-wide, principal-mode agrotechnologies can play any number of agroecological roles, but this should not eclipse their productive purposes. For example, a system can still be a windbreak for a neighboring agrotechnology, while productive output is the primary reason for use.

A number of agrotechnologies have been defined. These are

absorption zones,
agroforests,
aqua-agriculture,
aquaforestry,
alley cropping (hedgerow),
alley cropping (tree row),
entomo-systems,
forage (feed) systems,
intercropping (multi-seasonal),
intercropping (seasonal),
isolated tree,
microcatchments,
monoculture (perennial),
monoculture (seasonal),
parkland,
root support systems,
shade (heavy),
shade (light),
strip cropping,

support (perennial),
support (temporary),
terraces (constructed), and
terraces (progressive).

The topics in this chapter are condensed from various source descriptions (e.g., Nair, 1993; Wojtkowski, 1998, 2002). Given the recent recognition and evolving view of many agrotechnologies, this is not a well-developed topic. With the scarcity of examples, design packages are seldom fully understood. Without clear standard designs, notable variations are lacking, as many agrotechnologies have not reached this stage of maturation.

As an added note, some agrotechnologies overlap categories. For convenience, biomass banks and living fences are classified as auxiliary (described in Chapter 3) rather than principal-mode systems.

## *ABSORPTION ZONES*

Absorption zones are a water management system based on a depression (hole or ditch) immediately uphill (a few centimeters) from a perennial species. The hole is filled with loose organic materials, commonly leaves or small branches, with the goal of capturing and holding rainfall through saturation of the organic material.

The objective is to extend the growing season or increase the productivity for a target species. The additional moisture need not be retained over a prolonged period; a few weeks or even a few days can have a profound effect on increased productivity and decreased risk. This agrotechnology also increases the amount of nutrients available through decay of the added organic material.

### *The Standard Design*

One design is for widely spaced trees or shrubs. With this, micro-catchment channels can be employed to channel water to the plant, enhancing interception. A second design is for perennial hedges, again with an uphill hole or depression that is filled with organic materials. In this case, the depression extends along the upper side of the hedge.

## Notable Variations

With widely scattered trees, this variation may double as a parkland system, especially in the establishment phase. In dry hillside locations, it is effective for orchards with or without crops or for tree-based grazing.

Unmodified, this technique finds use in a labor-intensive tree establishment strategy where tree roots need moisture and time to reach a groundwater source. Alternatively, nonvegetative materials (e.g., starch formulations) are used to retain water in the root zone. These normally last for only one growing season while the roots become established.

## The Design Package

These systems are labor intensive and are best used where rainfall is scattered, intermittent, and intense, or where moisture is more prevalent at the start of a growing season than at the end. In some circumstances, drought resistance may be a desirable characteristic for the species used.

## Landscape Considerations

This system is best used in areas where groundwater sources are difficult to access or lacking, commonly on hilltops or areas with a shallow soil structure or where the groundwater is too deep for plants to reach in the establishment phase. These also serve as initial hilltop water control structures, beginning a series of coordinated erosion or water control defenses.

## AGROFORESTS

Agroforests are agricultural or forestry ecosystems containing mostly agricultural or potentially productive species, including a large percentage of trees. The design intent is to duplicate the natural dynamics (e.g., nutrient cycling, insect dynamics, etc.) of forest ecosystems. They can also be defined as an agroecosystem, comprising mainly woody perennials, based on high to moderate levels of plant density, species diversity, and spatial disarray.

## The Standard Design

Beyond density, diversity, and disarray, variations occur through species composition and system placement.

> *Homegardens*—These agroecological structures often surround houses where gaps in the trees induced by the house and other buildings can be an integral part of the light dynamics. In addition to production of ex-market food, spices, wood, and herbal medicines, these can serve an adornment function, produce a cooler microclimate around dwellings, and serve as an environmentally sound way to dispose of household organic waste.
>
> *Shrub gardens*—These are structures with the attributes of agroforests, but with short-stature perennials, a more open upper canopy, more light penetration at ground level, and more annual species. They also accommodate large populations of domesticated fruit species, those with fruit that is low branching and easy to pick.
>
> *Forest gardens*—These market-oriented agroforests generally produce a range of outputs, but with more emphasis on one or two primary species. They are generally located further from dwellings, stress marketable nonfood products, and have a greater percentage of woody output than other forms.

## Notable Variations

At the fringe of this category are natural ecosystems enriched with desirable species. This can include a forestry variation with the addition of (often more) high-value timber-producing trees or forests shifted more toward agriculture with additional fruit or other species.

Another variation is the temperate household garden. These lack much of the multistoried structure of their tropical counterparts, but still contain a high percentage of woody perennials and supply households with vegetables, fruits, nuts, herbal products, flowers, and other useful products.

## The Design Package

Agroforests are one of the more flexible systems, requiring only an area capable of supporting a forest ecosystem. The second requirement is a desire for small amounts of a large variety of outputs. Because of this, agroforests are seldom used for staple crops (notable exceptions occur in the South Pacific Islands with shade-resistant crops). Given the amount of in-use variation and the capacity for dual purpose, the socioeconomic side of the design package can be very accommodating.

## Landscape Considerations

Because of flexibility in use and an outward appearance as a forest fragment, agroforests are a common, if unnoticed, feature in most tropical landscapes. As a sole-purpose landscape entity, the home-garden version is often found around or close to households while forest gardens are more distant from higher-activity land use areas. As a dual-purpose structure, they can contribute to a positive landscape agroecology in any location.

## AQUA-AGRICULTURE

Plots that are flooded to raise specific crops are used in aqua-agriculture. Rice is the most common example; cranberries are another. Both these crops can also be raised outside a pond environment, but this requires abundant well-distributed rainfall, some form of irrigation, or naturally wet soils. As rice is the staple crop in many regions, aqua-agriculture is common.

## The Standard Design

To fill the crop's water needs, the standard design includes either a swamp environment or a pond situation created by supporting runoff or other irrigation. They are usually shallow and, for best results, the water level can be regulated.

*Notable Variations*

The Aztec floating gardens (mentioned in Chapter 3) were one extreme of this technology. Mounds for raising crops are used in flooded areas, as are parallel and closely spaced hillocks. In both cases, the water channels (1-2 m wide) hold heat, offer frost protection (see Frost, Chapter 7), and provide very moist soil conditions. Other variations have flooded trenches, but with a wider spacing, or use the banks of water canals (see Figure 6.2).

Because these sites are fertile and well watered, they are conducive to growing, without inundation, a wide range of high-value, high-yielding, nonstaple crops (e.g., vegetables). In addition to water-loving plants, flooded fields, canals, and channels can be used to raise fish as food or as a mosquito-control measure. Raised with crops, fish are a supplementary addition.

*The Design Package*

Pond-based agriculture is highly revenue oriented and finds use in agriculturally intense regions. These systems are expensive to establish, are species dependent, require large amounts of water, and need an accompanying water supply infrastructure.

These systems have the ability to bring otherwise unproductive and nutrient-rich swamps into production. Because water is not a limiting resource, they can have high yields with a corresponding reduction in per unit production and harvest costs. As such, they can also be a viable addition in regions where land use is less intense.

*Landscape Considerations*

The large volume of water required often necessitates a source that, given rainfall and topography, can require accompanying catchment systems, streamside locations, holding ponds, and/or wells. As such, the accompanying infrastructure can be a landscape feature impinging upon other land use practices. With their high maintenance costs, these can usurp management resources from other areas, forcing an economic landscape centered on aqua-agriculture.

# AQUAFORESTRY

With aquaforestry, trees are generally raised on dry land in close association with water. The primary crop is fish or other aquatic fauna. The trees have a facilitative role, providing (1) through shedding, insects, fruits, nuts, or leaf forage; and (2) through shading, cool water. This is also an example of agroecological mutualism, where trees benefit from the nutrients provided by fish populations and fish gain from the trees.

## The Standard Design

The standard design is based on the habitat needs of the fish (depth of a pond, water flow, temperature, etc.). In China, a wide variety of carp species consume green biomass. In Brazil, fruit- and nut-eating fish species exist and can be placed in a habitat designed for these species. In Chile, salmon needs determine the design, where willow trees, specifically weeping willow *(Salix babylonica),* above existing or diverted streams provide edible leaves and release insects into the water.

## Notable Variations

Ponds can be associated with agroforests. This usage is found in Southeast Asia, where cut-and-carry forage may also prove a viable option, necessitating a close proximity between ponds and biomass banks. This association is used for various species of carp in China (Gongfu, 1982). Gongfu also describes a variation where a fish, mulberry, silkworm association combines entomo- and aquaforestry.

## The Design Package

These systems have very specific water needs, fish requirements, and a correspondingly narrow design package. Despite this, these systems convert biomass into protein with comparatively little effort. The pond can be a visually pleasing addition to any landscape and, because of this, these systems can be highly desirable in the right location.

## Landscape Considerations

Some types of aquaforestry can require a major alteration in the natural landscape, while others simply take advantage of existing conditions. Aquaforestry can also be part of larger landscape scheme where holding ponds for aqua-agriculture are used for aquaforestry.


# ALLEY CROPPING

## Hedgerow

The basic design of hedgerow alley cropping has strips of seasonal crops raised between parallel rows of perennial hedges. There are a number of design uses, commonly including nutrient cycling, where the hedges provide green biomass for crops.

Equally or more important may be the potential for erosion control on steep, or less angular, hillsides. Hedgerows can also provide wind protection or serve in an insect control strategy.

### The Standard Design

The design has a primary crop, usually maize, with facilitative nonproductive hedges, only providing biomass output. The standard design has an interhedge distance of 4 m, where the hedges are generally pruned to a height of about 0.5 m.

Over 20 different in-use species have been mentioned; undoubtedly more exist. These provide plenty of opportunity to fine-tune the system. Normally, the hedges are a nitrogen-fixing, fast-growing tree or shrub species. Low-transpiration species may be best where rainfall is marginal. Decay-resistant species may be substituted in systems located on steep slopes, where soil or biomass accumulates around the base of the hedge.

### Notable Variations

Another version maintains the same design parameters as the standard design, but the hedges are trimmed slightly above ground level (Cooper et al., 1996). Mini-hedges facilitate mechanized agriculture with hedgerow systems.

Other designs can have a hedge as an understory to the primary species. Ruhigwa et al. (1994) describes hedges under plantains (*Musa* spp.) where the hedges add to the ground-level biomass. Another variation (Garrity, 1996) overcomes the permanence and high cost of removal for tree-based hedges, using instead a temporary hedge species, e.g., pigeon pea *(Cajanus cajan)*. The use of naturally occurring vegetation as a hedge follows the same lines.

## The Design Package

Because these systems have been intensely studied, more is known about the elements of the design package. These include

1. fertile soils without major nutrient limitations,
2. adequate rainfall during the cropping season, and
3. sloping land with an erosion hazard.

The socioeconomic part of this design package includes

1. an ample supply of labor,
2. high-intensity land usage, and
3. secure land tenure.

Further refinements are (Carter, 1996)

1. poor or declining soil fertility,
2. bimodal rainfall greater than 1,000 mm per year, and
3. soil pH greater than 5.5.

On the socioeconomic side:

1. maize as the primary crop,
2. high population pressure,
3. secure land tenure,
4. ownership and confinement of grazing animals, and
5. a need for firewood.

## Landscape Considerations

This system is generally revenue oriented and, without long-term productive and environmental benefit, is less attractive than a mono-culture of the primary crop. Thus, hedgerows are generally found in intensive land use situations where maintaining fertility and/or where erosion is a problem.

If the hedge is substantially taller than the crop, light interception and row orientation can be important (Ssekabembe et al., 1997) and dictates either location or management (i.e., more hedge pruning). For hillside locations, this will result in some design (agroecosystem or landscape layout) compromises.

## Tree Row

Tree row agrotechnology uses single, parallel rows of tall trees bordering crop strip rows. This is more complicated as the trees seldom have only a facilitative role, but may be an equal partner to, or the primary species for, whatever seasonal crop is grown. The key definable element is that the trees are touching within the row (intrarow tree spacing) and, because the area above the crop strip is open (the inter-row tree distance), direct sunlight can reach ground level.

### The Standard Design

The parameters of this technology are set by use. For commercial purposes, the standard design has a strip width sufficient to allow farm machinery to pass. A key element is the need for a small interspecies interface distance (between the tree and crop). This increases land use (LER) efficiency, but requires higher branching to allow machinery usage. The resulting clear stems increase the market value of wood-producing tree species, but for fruit and nut trees a higher than normal tree canopy height may increase harvest costs.

Row orientation to maximize light-use efficiency may be needed. This is generally north-south for maximum light apportionment where, in the morning and evening, horizontal light is allocated to the trees while noon light (vertical sunlight) is designated more to the understory crops.

Other orientations may be used to match crop light and/or water physiology with the intercropping environment. For example, where more moisture in the mornings, coupled with light availability, results in better yields, a row orientation that provides direct early morning light to the crops may be the better alternative.

*Notable Variations*

For many species, canopy spread may result in canopy closure. This condition may be a brief temporal phase in a long-term plantation sequence (see taungya variations in Chapter 5). Pruning can be used to delay closure or to maintain the long-term status quo. Other species, e.g., palms, which do not have spreading canopies, may be useful.

*Landscape Considerations*

Because row orientation needs may be topography dependent, erosion control may prove more important than light need. In this case, a different tree species or agrotechnology is contemplated (depending upon the value of primary and secondary species) or another topographic layout is used.

## ENTOMO-SYSTEMS

Insect-promoting systems are conceived to provide a food source and/or habitat for useful insects. They can be a stand-alone technology or integrated into other agrotechnologies. Auxiliary agrotechnologies can also double as entomo-systems.

### The Standard Design

These are specific-use systems where the standard design has a specific plant host species paired with the desired insect species. Examples are mulberries for silkworms, nectar sources for honeybees, or habitat for rare butterflies. Butterflies, with their specific vegetation needs, are raised and sold to collectors. Spacing is designed to maximize a food source and species pairing is designed to reduce predator insects.

### Notable Variations

A number of cultures include bugs as part of the diet (Menzel and D'Alvisio, 1998). Harvested insects often are a supplementary addition to a natural or productive ecosystem where, in systems not specifically designed to produce insects, they occur naturally and can be harvested. It is possible to increase the populations of these insects through system modification, e.g., adding more host species. Variations might involve providing predator habitat in an auxiliary setting for a landscape-wide predator-prey strategy. Honeybees can be a principle reason or a classic example of a supplementary addition.

### The Design Package

There is no standard design package except a desire for a specific insect species. This can include pollinating species or those that are part of larger insect control strategy (see Chapter 8).

### Landscape Considerations

These systems have dual-use flexibility and corresponding location flexibility. To be fully effective, the landscape cannot be at cross-purposes, where insect-suitable ecosystems are near those formulated to be insect inhospitable, e.g., used for insect control with principal-mode systems. If insecticides are employed, measures must be exercised to limit negative effects in insect-producing neighboring systems, e.g., barrier systems.

## FORAGE (FEED) SYSTEMS

Forage systems provide green biomass (forage) directly to grazing animals. Often these are simple pastures. More biodiverse systems, e.g., trees with grass, can serve a wider assortment of grazing animals (e.g., horses to graze trees, cattle eating the ground forage), produce more biomass per area, provide feed during long dry seasons when grass ceases to grow, and generally increase the use options.

### The Standard Design

Given the range of possibilities, there are a number of standard designs.

*Pastures*—These are often mixed, nonwoody, perennial grass species where forage is directly grazed. Other variations may be more monocultural.

*Trees with pasture*—These are pastures with the addition of woody trees or shrubs where both grasses and a woody component serve as a food source. The trees may be a dry-season source where, as grass growth slows, the animals eat the tree forage. In tropical regions, these are often formulated such that the more succulent grasses are consumed during the rainy season and the less palatable tree forage is the only available feed source during drier periods.

*Feed systems*—Another variation is a feed system where animals eat fruits, nuts, and other produce, rather than exclusively green forage. Such feed systems traditionally support pigs and, where climatically appropriate, rely on acorns (*Quercus* spp.). Aqua-forestry is the aquatic version of a feed system.

*Forage trees*—These are arid-zone systems where rainfall is too sparse to support grasses. Drought-resistant trees are substituted. In arid Chile, atriplex *(Atriplex nummularia)* is used in this role to graze goats.

## Notable Variations

A pasture can double as a cut-and-carry system with animals in another location. Another option incorporates living fencing in the design. The requirement is that the hedge species be complementary with the main forage species. With an inclusive living fence, the fence species can also be a forage source.

The use of animals to control various weeds fits under the heading of forage systems. Commonly, this is done to reduce fire hazard but, where the understory is detrimental to tree growth, animals can play a positive role. To be fully effective, this requires close attention to animal selection, timing (entry and duration), and stocking rate (e.g., Valderrábano and Torrono, 2000).

## The Design Package

Given the amount of variation, these are very flexible in-use systems. All that may be required are animals and forage (with monadic grazing) and, in more intense agricultural regions, fences to protect more valuable crops. The universality of forage systems is a testament to their flexibility.

### Landscape Considerations

As with the design and number of forage species available, forage systems can fit a wide variety of locations, either in labor-saving or labor-intense situations. They are especially useful as, with their perennial nature, they can double as riparian buffers and can play other ecological roles that extend outside their primary purpose.

## INTERCROPPING

### Multiseasonal

An intercrop that spans seasons has, by intent, perennial, often woody, component species. These can be short-lived, lasting only a few years, or long term, lasting many decades. The primary crop can be either the over- or understory, where all members produce useful outputs, which can include fruits, nuts, or any number of other tree based agricultural outputs or mixed forestry species. Facilitative multispecies versions also exist and are widely used.

Intercropping can describe any seasonal or long-term multispecies agriculture or forestry system. Multiseasonal intercropping depicts a long-term system where the components are density planted. This implies some degree of plant-plant complementarity.

### The Standard Design

A number of designs exist:

1. Mixed forest trees
2. Mixed orchard or tree crops
3. A combination of fruit-bearing and wood-producing species
4. Fruit-bearing and/or wood-producing with nonproductive facilitative species

This group overlaps with, but does not displace, other agrotechnologies. If the tree canopies are touching within, but not between, rows and the plant-plant interface distance is comparatively small, the design may be better categorized as a tree-row alley system. If tree canopies are high and completely overtop the crop rows, it can be considered a shade variant. Where this agrotechnology differs from others is that the canopies are more or less equal in height, light (or shade) is not a controlling mechanism, and all components are woody perennials.

> *Multispecies plantations*—Less utilized in forestry despite their productive and ecological advantages, these are a distant second to monocultural plantations in common usage. They can employ simultaneous planting, or the second species can be established after the first is in place.
> *Multispecies orchards/tree-crop plantations*—These are generally planted at the same time but, in contrast to their monocultural counterparts, many diverse perennial species are planted. These are not unusual, with many far-flung examples. The traditional streuobst of Germany are multispecies and/or multivarietal orchards with the option of a grazing or feed segment (Herzog and Oetmann, 2001).

## Notable Variations

Facilitative variations exist to promote the growth of primary species. These can have cover crop, shrub, or an understory or even overstory tree species to provide facilitative benefits to a productive species. Some forestry examples include, from the western United States, Douglas fir and red alder and, from Hawaii, *Eucalyptus saligna* with *Albizia falcataria* (Kelty, 1992). Komar et al. (1998) describe the substantially improved growth of teak when planted with *Leucaena*. Photo 9.3 shows an oil palm plantation in West Africa with a facilitative understory.

Also inclusive are multispecific or cross-species systems where, although the same class (genus) of plants is used, there is enough genetic variation to reduce risk. As examples, different species of pine, such as *Pinus resinosa* with *P. strobus* or *P. palustris* with *P. taeda* might coexist in the same plantation. As an agrotechnology category, this overlaps into perennial monocultures.

## The Design Package

Again, as with all agrotechnologies with large subgroupings, there is ample latitude to utilize intercropping in a number of situations. Where the perennial crops are hand harvested, labor is no more a factor than with monocultural plantations.

## Landscape Considerations

Multiseason intercrops can be, outside of staple crops, the most economically valuable of the cropping systems. In forestry form, as multispecies plantations, they can have more subsidiary or facilitative roles in a farm landscape.

## Seasonal

One characteristic of seasonal intercropping is that the component species are nonwoody and seasonal. A second is that a secondary species (one or more), because of relative height, can compete equitably with primary species for light. Seasonal intercropping can include different varieties of one species if, by intent, they accomplish agroecological objectives. They are often ecologically and economically superior to seasonal monocultures and are the focus of much of agronomic agroecology.

## The Standard Design

A number of standard designs exist. It is difficult to say which version is more common. The key aspect is the degree of plant-plant complementarity.

>   *Facilitative*—In the facilitative version, the species that accompanies the primary species serves a facilitative purpose. A common example is the use of a cover crop with a primary species; numerous variations exist.

*Multioutput*—With this variation, each component species produces some output and, in many cases, there is no clear primary
species. Maize with beans is a common biculture, examples of
which are found widely in the Americas. In triculture form,
beans, maize, and squash coexist successfully. A partial list of
common intercrops, numbering about 55 pairings, has been
compiled by Vandermeer (1989, p. 2).

## Notable Variations

Often observed are low-density intercrops, where only a few
widely spaced secondary plants are found among a normally spaced
primary species. For example, maize with potatoes may have the
maize with a 4 to 6 m interspecific spacing and potato at the normal
density. The maize is hand picked before the potatoes, which can be
machine or hand harvested.

A second variation is a partial suppression design where, with high
rainfall, both thrive. With low rainfall, one dominates while the other
is excluded. With this system, a land user overcomes some of the vagaries of the climate.

Also possible is multispecific or cross-species intercropping. Although the same class (genus) of plants is used, providing the same
product output, enough genetic variation remains to resist some natural stresses. This may also be considered a monocultural agrotechnology.

## The Design Package

There are limitations in mechanized agriculture in that harvesting
and separating two intertangled crops is problematic. Other obstacles
include knowledge of the intercropping possibilities.

## Landscape Considerations

As a means to produce staple or high-value market crops, seasonal
intercrops can be the economic center of the landscape. Any subsequent systems may be placed to protect or support these systems.

## ISOLATED TREE

Isolated tree systems are large areas containing one or two scattered trees. Since the trees cover less than 5 percent of land area, overall ecological gains may be relatively small, but sufficient for widespread adoption. In practice, the reasons for acceptance may be secondary productivity gains, e.g., promoting bird and bat habitat for insect control, to provide a shady rest for workers and animals, and/or as a place to store forage above hungry animals (i.e., horqueta trees).

### The Standard Design

Except for wide spacing and a minuscule tree planting density, there are no requirements or standard designs. In contrast to parkland systems, the trees can be of any species and possess a wide variety of characteristics. Harvey et al. (1999) noted 15 species used in this role in the pastures of Central America.

### Notable Variations

In many uses, the trees serve only as a source of secondary outputs, to increase biodiversity and, through their presence, attract insects and fauna (Harvey et al., 1999). Although most noticeable with seasonal crops and in pastures, isolated trees can be found in many short-statured polycultures or even mixed within tree crop or forestry plantations.

### The Design Package

As the trees contribute or subtract little from the primary crop, they share much the same design package as monoculture or pasture systems. The additional gains, although small, may be enough to encourage use.

### Landscape Considerations

As a system formulated to produce staple or market crops, isolated tree systems can be prominent in farm landscapes. Within a small area, the ecological role of these trees is minor. In larger areas, this role could increase enough so that the total effect can be significant.

# MICROCATCHMENTS

Channels designed to funnel water to single shrub or tree species are microcatchments (see Photo 4.1). These are usually found in semi-arid regions. Catchments can be permanent or temporary structures. The latter are used only until tree roots reach a groundwater source. They can be a long-term feature in dryland principal-mode forestry, tree crop plantations, or orchards.

## The Standard Design

The standard design for microcatchments are V-shaped channels, usually 0.5 to 1 m in length, facing downslope with one (less commonly more than one) perennial species at the apex of the furrow. In size, they should capture enough rainfall to be effective, but not enough to overwhelm the plant.

Farrow variations are possible, where rows of plants share a single trough. Again, the amount of water captured and channeled to a tree or hedge determines spacing.

## Notable Variations

A simpler version is a checkerboard pattern of shallow depressions. Another uses the microcatchment as part of an absorption zone. As a separate agrotechnology, the latter is described at the beginning of this chapter.

## The Design Package

Design packages are revenue oriented and may be part of an overall water defense. Where temporary, they are employed in the first stages of tree planting when survival is critical. When microcatchments were employed, Gupta et al. (2000) found tree survival to improve from 50 to 90 percent, but planting costs increased 20 to 30 percent.

## Landscape Considerations

Catchments, along with microclimate control, allow trees to be planted where survival is difficult. They expand the landscape possi-

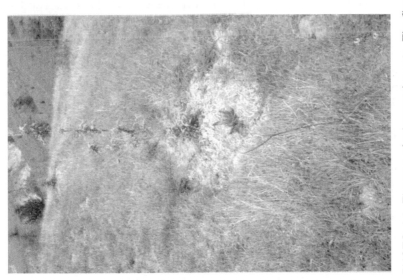

PHOTO 4.1. The use of microcatchments. The first photo (left) shows a single microcatchment, the second (right) shows a field of microcatchments in use to establish a block of trees.

bilities in arid or semiarid zones. Tree spacing, slope, and rainfall patterns dictate where and how catchments are used.

## MONOCULTURE

### Perennial

Single-species systems with an extended life are found in agriculture or forestry. The more revenue-oriented systems (tree crops) can tolerate less unintended biodiversity (the occurring agroecosystem), while the cost-oriented systems (usually forestry) contain considerable biodiversity in the form of fauna and small, often unintended or unwanted, plants (i.e., weeds).

### The Standard Design

Perennial monocultures can be divided into forestry and tree crop plantations, although other plants (e.g., cacti or bamboo) qualify. Examples of tree crops are fruit trees (i.e., orchards), palms (oil, coconut, etc.), or other crop-yielding woody perennials such as rubber trees.

When plants remain in place for a long duration and seasonal outputs are kept within set limits, these systems are not exposed to many sustainability dangers. If they are, measures can be taken to mitigate any sustainability and/or ecological stress to which a particular species may be susceptible.

### Notable Variations

The notable variations often involve unplanned flora and fauna additions that are tolerated, contribute either positively or negatively to the overall ecological impact, and provide alternative or supplementary products. These can be at ground level or in the canopy.

Multivarietal plantations, those that keep within the same species class (genus) and output type, can provide genetic variation with associated protection. This is an unstudied option that may not be necessary where more biodiversity, as with mixed-species plantations, does not incur any additional costs, e.g., at harvest.

## The Design Package

Design packages divide into high-intensity systems with seasonal or continual outputs (oil palms, coconut palms, and rubber trees are examples of species that continually produce) or low-intensity forestry plantations with one final cutting and harvest at the end of the cycle.

## Landscape Considerations

The simplicity of perennial monocultures allows for considerable latitude to utilize landscape considerations in variety of principal-mode and auxiliary roles. Catchments, windbreaks, and shelterbelts for animals represent a few of the many landscape options.

## Seasonal

Monocultures of only one species are often of short duration. They serve as a benchmark and, because of common usage and ease of study, are a basis for both productive and economic comparison (e.g., LER and RVT).

The advantages include one planting and one harvest, with little agronomic complexity. The disadvantages are that monocultures are more susceptible to the different forms of risk, can lack sustainability, and, in terms of LER, are not overly efficient with site resources.

The simplicity of monocultures makes them well liked by lending groups (e.g., banks), as it is easier to monitor and evaluate performance (Godoy and Bennett, 1991). This bias is more pronounced with larger commercial farms and may contribute to their overwhelming use in these situations.

## The Standard Design

With comparatively few variables to consider, some of which are dictated by the farm situation, designs are easy to implement. For example, wheel spacing on tractors can set the interrow spacing, or harvest and planting machinery can set the number of rows per area.

The most ecologically fundamental monoculture is clonal, having no niche variation between the component plants. The lack of variation is the result of a single parent plant being the source of the offspring. These are becoming more common. In most seed-planted areas there is some, often minor, genetic difference and minor niche variation between the component species.

## Notable Variations

In unadulterated form, monocultures have only one species. A varietal version is based on niche and genetic diversity. In many commercial species, there can be considerable genetic diversity between plant varieties, and this is used to gain the advantages of biodiversity while maintaining like outputs.

Multivarietal rice plantings have been used to reduce disease loss in China (Yoon, 2000) and in mountain-grown Pakistani wheat (MacDonald, 1998). The amount of cultivar variation can be extensive. Boster (1983) found 15 cultivars of cassava in one tropical farm, which accounted for a considerable percentage of the overall agrobiodiversity.

Cross-varietal monocultures can include different species within the same genus, e.g., mixing the bean species *Phaseolus vulgaris* and *P. coccineus*. These have like outputs, but retain some of the advantages of biodiversity. This form of biodiversity might also be categorized as seasonal intercropping.

## The Design Package

These systems are the most common cropping methods and, because of their simplicity, are heavily favored across cultures, land types, and other variables. The multivarietal versions, although mitigating some of the ecological risk inherent in less genetically diverse systems, are far less understood and used.

## Landscape Considerations

Monocultures are prominent features in most landscapes. Because these systems have the highest ecological risk, they require high maintenance with regard to sustainability, landscape layout, and long-term planning.

## PARKLAND

Parkland systems have scattered trees within farm plots. From 5 to 50 percent of the plot area is covered by trees. In classification terms, these systems lie between scattered tree and either tree row or light shade systems with almost continual cover.

Environmentally, the trees serve a number of purposes. Some nutrient gains are possible (e.g., Belsky, 1992; Belsky and Canham, 1994; Boffa, 2000), but crop yield losses may be more common. The trees can be used to store forage, as shade for cattle, and aid in insect and rodent control (see Chapters 7 and 8). The crop or pasture is the primary species, while the trees could produce a secondary output or be a wood source.

### The Standard Design

The standard design is regionally based, utilizing a common species. Examples are baobab *(Adansonia digitata)* and *Faidherbia albida* (parts of Africa), *Prosopis cineraria* (parts of India), *Nothofagus obliqua* (southern Chile), and species of oak (southern United States, e.g., Bainbridge, 1988). Both *Faidherbia albida* and *Prosopis cineraria* have been shown to have a positive effect on crop yields (Sanchez, 1995; Young, 1989b, pp. 161-167).

### Notable Variations

With the exception of tree and crop species selection and tree planting density, there is little design flexibility with parkland systems. The most notable variation is where trees have less of a facilitative role, instead providing a second product at the expense of some crop yield reduction.

A number of these variations exist. In Africa, the baobab *(Adansonia digitata)* and shea trees *(Vitellaria paradoxa)* provide a secondary output (respectively, Sidibé et al., 1996; Boffa et al., 1996), while in southern Chile, sweet cherry *(Prunus avium)* fills this role.

### The Design Package

The only requirement is that a suitable species exist and that the trees provide enough benefit to overcome any associated costs or yield reductions.

## Landscape Considerations

Parkland species can be an element in a landscape-wide insect and rodent control strategy (see Chapter 8). They may also be part of an integrated layout, where some forms of protection are extended through the parkland design.

# ROOT SUPPORT SYSTEMS

Root support systems are used where the primary species is subject to lodging or toppling, which can result from strong winds, shallow roots, shallow soil, and/or a loose soil structure. Toppling may also result from harvests when ladders and/or machines push over trees.

## The Standard Design

Supporting vegetation can be used in orchard or tree crop plantations as a closely planted understory or overstory. The desirable characteristic of the supporting plant is a high degree of plant-plant complementarity and strong spreading roots. As an understory component, this need not interfere with harvests. As overstory, these should have light complementarity with the primary species.

## Notable Variations

The use of support is not restricted to perennial woody species. Annual crops do suffer from lodging and resulting loss of yields. A suitable cover crop may serve not to anchor roots, but to prevent stalks from bending. For this purpose, the cover crop should be dense at a height not to cover the leaves, but to support the stem.

For perennials, root support may be secondary in a larger list of ecological benefits. Nutrient facilitation may top this list, where insect, wind control, and water management are also among the benefits gained.

## The Design Package

Root support is utilized with high-value crops, mainly in orchard or plantation situations. The danger or secondary purposes must be important enough to adopt this drastic approach, rather than using less severe alternatives such as windbreaks. Generally, it is used in revenue-oriented systems.

## Landscape Considerations

Root support is unusual and is principally found only where other measures (e.g., classic windbreaks) are not feasible. This may occur where land pressures are intense or the agroecosystems too small or too exposed to winds to merit wider landscape defenses.

# STRIP CROPPING

This system alternates strips of different crops, crops and fallows, crops and grasslands, or crops with other vegetation. These are formulated so that only alternating or scattered strips (but not adjacent) are exposed to the same ecological danger. As such, these strips moderate various forms of natural stress.

As an agroecological addition, strip cropping is primarily a counter against erosion (wind or water) and an insect control addition. Taller strips may find use in protecting animals and crops from weather extremes.

## The Standard Design

The standard design has alternating cropping systems using different species. These are best when season crops are paired with longer-term, nonwoody or woody perennials.

With sufficient width, strips lend themselves to machine use, and the dimensions are usually dictated by the width of plowing or mowing attachments. Most often, the strips run along slope contours, although some counter-contour placements are possible (see Chapter 6, Figure 6.1).

## Notable Variations

As a cropping-fallow system, strips need not be permanent, but be part of the fallow sequence. Each strip can be one phase in a rotational sequence where a fallow, productive or nonproductive, is included (see fallows, Chapter 5).

Similarly, strips may be permanent and/or part of a cut-and-carry system. The permanent strips may contribute to nutrient dynamics of crop strips directly, through aboveground leaf fall or crop roots extending into the nutrient-rich strip soils (Wijesinghe and Hutchings, 1999; Farley and Fitter, 1999).

Cut-and-carry strips expand upon this. To reduce labor needs, they may utilize a counter-pattern where rows within the cut-and-carry strip run perpendicular to the strip orientation. In-use examples are provided by Versteeg et al. (1998).

Another variation has strips of permanent taller trees alternating with crop strips. One advantage of trees over strips of short-statured species is in frost protection (Wang, 1994). This shelterbelt design is used on a larger scale (with wider strips) to produce timber, moderate temperature, and protect animals grazing in the pasture strips in or between the trees (Moore and Bird, 1997).

## *The Design Package*

As the easiest cropping system (other than the monoculture) to implement, strip cropping has wide potential to address a range of sustainability and risk issues. The only caveat is that sufficient land area must exist and the climate and topography be suitable.

Because strips can be used with farm machinery, the various strip designs have wide application. This extends to forest plantations where, as harvesting exposes land to degradation, strips provide erosion control and other ecological benefits.

Strips find use on organic farms. The strips provide for a reserve of predator insects or corridors for their movement, helping to keep crop herbivore insects at bay. Strips also may be part of a strategy for reducing chemical fertilizer inputs.

## *Landscape Considerations*

Strip cropping is used on slopes that are not overly steep (<25 degrees). Even mild grades (2 to 10 degrees) benefit from strips when there are loose soils and an erosion danger. In all these cases, light and water needs may dictate use and placement.

## SHADE SYSTEMS

### Heavy Shade

These systems are characterized by dense, total canopy coverage and a high degree of ground-level shade. Along with nutrient cycling, the advantages of heavy shade are in insect control, eliminating drought-induced stress through below-canopy microclimate control, and weed reductions. Normally, there is a tradeoff between potential yields, reduced risk, and reduced operating costs, which makes these systems more cost oriented.

### The Standard Design

The most common design uses a single tree, usually nitrogen fixing, above one crop species. Shade-resistant species are conventionally used in this role, e.g., coffee, cocoa, vanilla, or black pepper.

To be fully effective, these must cover a fairly large area, or any light input through edges or gaps should be eliminated through use of separate shade and border species. Shade is generally regulated through species selection (i.e., the use of leaf area index to measure the amount of shade) and, to a lesser degree, through management (pruning or thinning).

### Notable Variations

Variations include the use of natural forest (mixed species) canopies or planted multispecies canopies. Some can be fairly biodiverse. In Côte d'Ivoire, of the 27 wild tree species used as shade above cocoa, 13 provide firewood and medicine, 11 provide food products, and 6 are used in construction (Rice and Greenburg, 2000). Having the same upper canopy potential, pasture systems can also be shade systems with shade-resistant grasses.

### The Design Package

As cost-oriented systems, heavy shade is usually employed in situations where labor and/or other inputs are restricted. These systems usually do not use farm machinery, so the understories are most suitable for pasture or the production of handpicked fruits or nuts.

## Landscape Considerations

As a landscape factor, the reduced cost aspect permits a larger area to be farmed less intensively. The tradeoff is lower output and returns. The result is that these systems find more use in regions with low land use intensity. With predator-prey dynamics, the canopy provides overhead reservoirs and travel corridors for predator insects and habitat for insect-eating birds. Insecticide use near these systems would not be helpful. With good overall stress management properties, shade has extended landscape possibilities.

### *Light Shade*

Light shade systems have an overstory, but either widely spaced or with an open canopy to permit ample light to reach understory crops (see Photo 4.2). Generally, the canopy covers 80 to 100 percent of the

PHOTO 4.2. A light shade system with unidentified component species. In this example, the open canopy and comparatively high degree of understory light are conducive to crop growth.

land area. The overstory can be a nonproductive facilitative species, but productive species can be used.

In contrast to heavy shade systems, species selection and complementarity is more of a factor. For example, Faizool and Ramjohn (1995) suggest that nitrogen exchange is a positive influence in cacao yields under a light shade canopy.

Light shade systems differ from agroforests in their ordered structure, lower biodiversity, and their reliance on plant-plant dynamics. They differ from multiseasonal intercrops in that shade and light are key factors in regulating overall growth, with less reliance upon plant-plant complementarity.

## The Standard Design

These are revenue-oriented systems with considerable flexibility in use. Normally, all components produce some output. Examples include coffee with various mixes of banana, shade, fruit, and timber trees (Escalante, 1995; Ashley, 1986).

## Notable Variations

Usually the understory contains the primary species, although, through a reversal of standard design, the overstory contains a facilitative species designed to improve nutrient cycling for the overstory. Coconuts above the tree species gliricida are one such example (Liyanage, 1993).

As biodiversity increases, these systems begin to resemble open-canopied agroforests. As overstory density increases, they become variations of heavy shade systems.

## The Design Package

Light shade finds use as intense systems where two, three, or more outputs are desired from perennial plants. These can include staples or crops of high market value where, through light management, the mix of outputs is adjusted.

## Landscape Considerations

As a well-protected perennial system, light canopies have wide ecological possibilities in the overall landscape. Because they are more revenue intense, they can economically impact farm dynamics more than heavy shade, and this is often a selection criterion.

## SUPPORT

### Perennial

Support systems utilize perennial wood species to support perennial vine crops (see Figure 4.1) Examples include grapes, hops, kiwis, vanilla, passion fruit, and black pepper. For classification purposes, a single vine crop with an artificial trellis is considered a monoculture.

### The Standard Design

The standard design includes vines grown over or within the canopy of the supporting plant. A number of supporting species will assume an umbrella shape upon pruning, and vines can be grown over these. A number of temperate species have this property, e.g., Scotch elm (*Ulmus glabra* var. *horizontalis*), the weeping higan cherry (*Prunus subhirtella* var. *pendula*), and weeping beech (*Fagus sylvativa*

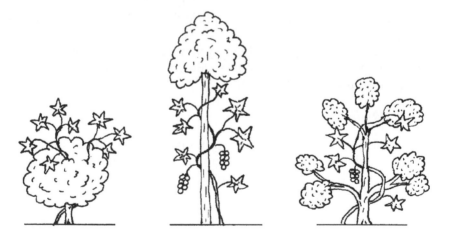

FIGURE 4.1. The three types of perennial support, from left to right, are (1) vine over tree, (2) vine under canopy, and (3) vine in canopy.

var. *pendula*). Also used are trees where severe branch pruning permits vine growth within the canopy, e.g., FAO (1994). Living support provides for nutrient cycling, insect control, and other ecological advantages.

## Notable Variations

Another variation (Salam et al., 1991) has the vine growing on the tree stem below the canopy. Considerations for this tree-over-vine alternative involve spacing and a tree species that will allow sufficient light to permit acceptable vine crop productivity. This formulation interferes less with tree growth and yields and can relegate the vine crop to a secondary role.

## The Design Package

Vines generally require hand labor, so mechanization factors are a smaller consideration. The tradeoff is with expensive, but maintenance-free trellises versus the cheaper, higher-maintenance natural support. A factor to consider is the ecological gains from the planned biodiversity of support systems.

## Landscape Considerations

Natural supports are an option whenever vines are grown. When vines with trellises manifest stress problems, the ecological counters include living supports and/or using neighboring principal-mode or auxiliary systems.

## Temporary

Temporary support systems use fast-growing woody species to support a short-term vine crop. Often the trees are killed or suppressed while the vine crop is being raised.

## The Standard Design

The most common system has vines growing over a living tree canopy. Any number of annual or short-lived vine species can be used, perennial or annual, e.g., yams, climbing beans, and cucumbers. The trees can be young, short statured, or maintained in a diminutive form through pruning.

## Notable Variations

There are variations of the standard design in which dead stems of newly raised trees are used to raise annual vine crops (e.g., Rachie, 1983). Three options exist for this: (1) the trees may raised during the crop fallow and killed at the beginning of the cropping season; (2) the trees are raised in rotation, where one row or strip is used for tree establishment and the other for support; and (3) the tree stem is killed by girdling and the rootstock produces new stems in a subsequent growing season.

## The Design Package

This system can be cost-oriented (replacing expensive trellises) for use on nonmechanized farms. The option for firewood expands its attractiveness in subsistence situations.

## Landscape Considerations

In addition to the strip crop possibilities, the tree-vine combination can possess some nutritional attributes that would make these systems suitable as part of a rotational sequence.

## TERRACES

### Constructed

Constructed terraces come in various forms, where the purpose is to provide a level surface for cultivation, while increasing water retention (e.g., with rice paddies) or infiltration (e.g., with dryland crops where rainfall is brief and intense).

### The Standard Design

There are a number of terrace options, but the standard is an earthen terrace (see Photo 4.3). These are dug out of hillsides and the steep slope is covered with grasses or a cover crop.

PHOTO 4.3. This series of earthen terraces, as used in a high-intensity application, is a demonstration site seen in the early spring before crops are in place.

Trees may also be added to increase the productive potential by utilizing the face (sloping part) of the terrace. For this, the trees are usually planted on the lower third of the terrace face. These trees are short statured or pruned such that canopy does not overly shade terrace crops.

*Notable Variations*

There are two variations, the first an uncommon variant of the earthen design, the second more widely used.

> *Buried hedge terraces*—This terrace is formed by burying a pruned hedge. This provides a structural base in the more erosion-prone early period, while grasses or other vegetation are being established or the original hedge re-emerges.
> *Stone terraces*—These have use where stones are plentiful near the site of a proposed terrace. They can be stone only or

strengthened through the use of trees (FAO, 1994). They have the additional advantage of being able to store daytime heat, reducing nighttime frost risk (MacDonald, 1998).

*The Design Package*

Terraces are generally found in hilly regions of very high agricultural intensity. They are an economically and ecologically prominent feature in any landscape, as considerable effort is expended in maintenance. This is reflected in the crops raised (short term, higher value) and the continued cropping needed to recoup the effort expended. Because of cost, they have few, if any, forestry uses.

*Landscape Considerations*

Terraces convert less agriculturally promising hillsides into high-intensity sites. They are found where agriculture must expand to high-risk areas.

## Progressive

Progressive terrace designs are less common, but are finding use where increased land pressures force more intense farming on steeper hillsides and where, because of land steepness, less extreme antierosion measures (strips) will not suffice. This form of terrace is far less labor intensive than constructed terraces and may find favor where fallow periods are utilized. During this period, the hedge continues to grow while the accompanying vegetation will capture and store nutrients.

*The Standard Design*

The standard design uses contour rows of trees, which, once established, are cut to a convenient height, and cut branches, along with other debris, are piled on the uphill side of the tree row. Erosion, often promoted by plowing, begins to slowly form a terrace. It can take a few years before the process is complete (Banda et al., 1994).

*Notable Variations*

Rather than a single row of trees, a closely spaced (less than 0.25 m) double row may be substituted. This provides a greater amount of biomass, slightly speeds the terrace-forming process, and supplies more nutrients.

## The Design Package

These terraces require land pressures such that farmers are willing, or need, to farm high-erosion, high-risk landscapes. This design package is used where labor availability and fallow periods may not be conducive to fully constructed terraces, and the time frame involved favors a slower, less intense approach. These terraces are also an alternative where expensive inputs (fertilizers) are beyond the means of farmers.

## Landscape Considerations

Progressive terraces find use in the same topographic situations where constructed terraces are found. Because they are generally used in conjunction with a long fallow, a larger area is needed than for terraces implemented without the fallow period.

# Chapter 5

# Temporal and Auxiliary Agrotechnologies

This chapter looks at two groups of agrotechnologies. The first are temporal and describe, over a set time period, planned growth sequences. These are denoted with the letter T. The second group is the auxiliary agrotechnologies. In contrast to principal-mode systems, they do not have production as primary objective, but instead confer other agrotechnological properties (i.e., ecological or economic services) on neighboring systems. Agrotechnologies in this group are designated by an A.

As mentioned in Chapter 3, temporal sequences are part of the description of a principal-mode system, but over the planned progression, a temporal agrotechnology can embody any number of principal-mode agrotechnologies. The key detail in defining a temporal agrotechnology is not the length of the cropping cycle, but the planning involved.

If one cropping system is to follow another for economic and/or agroecological reasons over single or multiple seasons (i.e., crops or trees), then it is a single agrotechnological sequence. If there is no overall design or planning in the sequence (i.e., the land user does not know what will follow or there is no direct economic or planned agroecological connection between the different crops), then each cropping phase and temporal agrotechnology begins anew.

A second group, the auxiliary agrotechnologies, exist primarily as a facilitative aid to a neighboring principal-mode agrotechnology (e.g., windbreak) or serve some other ecological, nonproductive landscape function. These are entirely subordinate structures with regard to any useful output. As such, an auxiliary agrotechnology can replace or reinforce any number of landscape ecological functions, except production capacity.

Landscape considerations for auxiliary systems associated with water, wind, and insect control are not listed. They are presented in Chapters 6, 7, and 8.

## SOLE CROPPING (T)

The sole crop is a system in which one principal-mode agrotechnology exists on a given area and the end of the sequence terminates the temporal phase, or where there is a planned sequence of the same species. Sole crops are exclusive to those agrotechnologies where all component species (one or more) are of comparable growth duration.

### The Standard Design

The agrotechnological period can be single or multiseasonal, commonly having a single species, but polycultures of comparable-duration species are possible. These systems can use agricultural crops or span many years, as with a forestry or tree crop plantation, but with the proviso that there be no planned ecological connection between past, present, and future land uses. To maintain acceptable yields, imported nutrients and other resources are often required in high-output situations.

As with monocultures, sole cropping serves as the standard of comparison. This mostly involves sustainability questions.

### Notable Variations

As the most fundamental temporal pattern, simplicity limits the number of variations. These systems are not always monocultural. The crops or tree crops can have accompanying vegetation whose growth duration does not fully coincide with that of the primary species. For example, with a maize and bean biculture, the bean component may be planted a few weeks after the maize.

### The Design Package

This type of package is known for flexibility, simplicity (especially in monoculture form), and, with crops, the short time frame involved. The longer-term sole-crop systems, those with plantations or orchards, require longer time horizons and a greater landscape commitment.

## Landscape Considerations

As the most common temporal sequences are often associated with staple crops, these systems often occupy the best farm sites. In high-volume, high-input form, these crops must be close to the transportation system (roads, etc.) and easily accessible to a land user. Without internal measures to protect the soil and crops from the vagaries of nature (especially high-exposure seasonal crops), these sites should be safeguarded through appropriate landscape ecological associations.

Long-term forestry rotations are more stable and can be positioned on high-risk sites. Multispecies versions, those with natural stress resistance, are also useful from a landscape ecological perspective.

## ROTATIONS (T)

Rotations involve a planned change in nontemporal agrotechnologies, species, and/or plant varieties over time. With rotations, there is no overlap between phases; when one ends, another follows. For classification purposes, a fallow may be part of a series of rotations. These are used in both agronomy and forestry and are a distinct course of action in landscape agroecology.

As part of a single design sequence, rotations allow for more efficient nutrient use, e.g., a nitrogen-fixing crop may be succeeded by a nitrogen-demanding species. Rotations may also be used to disrupt insect reproduction cycles and as a weed control measure.

## The Standard Design

This system has a clear standard design in which one seasonal monoculture (or intercrop) is followed by another in a predetermined sequence. The variation comes through the species used in each phase. Any number of crop species can follow in sequence, including multiseasonal species or seasonal intercrops.

The general idea is that one species sets the nutritional stage for the next species in the sequence. In a seasonal, nutrient-facilitative rotation, the crops are often capable of being intercropped, but because water is the limiting resource, sequential cropping is the better alternative.

## Notable Variations

Climatic aspects may determine a sequential ordering. These may follow rainfall cycles where uneven bimodal rainfall (two rainy seasons in a single year) exists. In this case, a higher-value, water-demanding crop is raised in the wetter season. A nutrient-compatible, low-value, and less water-demanding species is produced during the dryer phase. The remainder of the year constitutes a brief fallow.

The predetermined sequence may be the most common but not the only rotational strategy. The alternative is a plug-in approach where the temporal attributes (nutritional needs and resulting postcrop soil conditions) for each crop species are known and, on this basis, a mix-and-match rotation is undertaken. For this, sample guidelines are devised and these serve in formulating an order.

The case study at the end of Chapter 10 samples some of these crop-based guidelines. This alternative offers a bit more cropping flexibility than a predetermined sequence, but requires more research input.

## The Design Package

Rotations have a number of uses, and these are reflected through the design package. If soil nutrient requirements are addressed through rotations, a sufficient land area and a value for each cropping sequence (e.g., markets that can absorb the production from each phase) must allow this to take place. Rotations may also serve as a temporal barrier to the spread of herbivore insects and/or plant diseases.

## Landscape Considerations

Any rotational sequence must be accommodated within the larger landscape, requiring enough land for an economically viable harvest of each crop in the planned sequence. Soil characteristics are less important, as a sequence is designed to overcome less severe nutrient limitations.

With some planning, rotations have less exposure to natural stresses than sole cropping and, to reduce these stresses further, must be safeguarded across the wider landscape. This especially includes high erosion risk from a seasonally exposed, bare-ground phase.

## FALLOWS (T)

The use of fallows within a cropping sequence is standard practice in many regions. Given the high nutrient removal associated with crops and low nutrient demands of tree plantations (Fox, 2000), fallows are exclusively associated with seasonal or other short-term, high-output systems not in forestry situations.

The goal of the fallow is to regenerate soils such that per area yields are above, and per unit harvest costs for subsequent crops are below, an acceptable limit. Fallows can be inserted between a series of individual rotations. Fallows constitute a separate temporal agrotechnology when there is no plan for the succeeding ecosystem. Normally, this occurs with long fallows. Because yields are secondary, fallows can also be classified as the only transitory auxiliary system.

### Standard Design

Fallows subdivide into purely facilitative and longer-term productive fallows, which allow for low levels of outputs. Generally, these categorize well, but with some exceptions.

### Facilitative

If there is a standard design, it is the use of a woody, burned multispecies fallow where naturally occurring vegetation is cut and burned. The fallow periods range from one season to over 50 years, land area permitting. These practices are found worldwide in low-intensity agricultural zones.

More recent innovations have included the use of high-biomass, fast-growing species that accumulate nutrients quicker and permit expanded cropping. The species can be a woody and/or nonwoody.

*Productive*

With the common exception of grazing, fallows generally produce little economic value. The alternative is a productive fallow where, by design, a sequence of outputs is generated. The idea behind a productive fallow is to enter a planned fallow sequence to rebuild soils, while producing low-value and low-yielding outputs.

In many aspects, this is similar to a rotational sequence, except that each subsequent stage of the sequence has longer-term and larger-statured perennials than the previous stage, and there can be considerable overlap between stages. The ending phase can be an enriched forest or an agroforest or any point in this progression.

Since a fallow usually starts with depleted soils, nutrient gains come through species diversity, only a few of which are productive. This is often coupled with extensive use of the nonharvest option, where only a small segment of the productive output is removed, and the remainder, usually of lower quality, recycled.

*Notable Variations*

Described here are some of the documented variations on facilitative fallows. These are based upon whether the fallows include woody or nonwoody vegetation, are burned or not, and are monocultural or polycultural. This gives eight variations ($2 \times 2 \times 2$).

The first of these, a woody, monoculture fallow, is found in southern Brazil using the tree species bracatinga *(Minosa scabrella)*. The trees are cut and burned and the crops planted. Fire germinates the tree seed, and resulting trees are thinned when the crops are weeded. The fallow reestablishes at the end of the cropping season. A second example comes from Indonesia (Christanty et al., 1997), where bamboo is the fallow species. This is cut and burned. The crops are planted and, following a two-year cropping phase, the bamboo regerminates. After a four- to five-year fallow, the sequence begins anew.

In the highlands of Costa Rica, pure stands of caragra *(Lippia torresii)* are harvested for firewood while the decaying green biomass provides crop nutrients. The trees later reestablish from stump sprouts.

Even less common are woody, unburned, multispecies fallows. The sequence is similar to the caragra case, but reestablishment from seeds or stump sprouts depends on the species (Kass and Somarriba, 1999).

For the nonwoody, burned monoculture fallow, specific plant species are required. Burning is needed to allow time for the crops to grow without being overrun by cover reestablishment. Fire can germinate the seed or delay sprouting depending on the fallow species used.

A nonwoody, burned multispecies fallow is, in most cases, based on mixed grass species, where the purpose is not to kill the grasses but to delay regrowth. An example from Mexico (Gliessman, 1998, p. 76) is unusual in that burning occurs immediately after a maize crop is planted. The seeds are protected from heat by deep planting and soil moisture.

There are examples of nonwoody, unburned monoculture fallows (Kass and Somarriba, 1999). Canavalia *(Canavalia ensiformis)* and lablab bean *(Dolichos lablab)* are used in drier areas of Latin America. These cover crops are cut to ground level and maize is sown. The regrowth of the cover crop (which is slower than the maize germination) eliminates weeds, and, at the end of a single maize crop, regrowth reinitiates the fallow. A nonwoody, unburned polycultural fallow is similar to burned fallow, but with natural decay instead of fire to release nutrients.

### The Design Package

The examples in the previous section exhibit a lot of variation. The essential component is the need to replenish soil nutrients using vegetative sources. This need has generally kept these systems within the sphere of staple crop subsistence farming, but with demonstrated potential for commercial farming (e.g., Jordan, Hutcheon, and Donaldson, 1997; Jordan, Hutcheon, Donaldson, and Farmer, 1997). Another element is sufficient land area to support cropping given the length of a fallow and the area utilized.

### Landscape Considerations

The use of a facilitative fallow does allow for flexibility of landscape layout but, unless the fallows are short, they are not favorable where extensive and expensive land infrastructure is required (irrigation, better roads, etc.). On a small scale or in league with some agrotechnologies (e.g., strips), there is the option to use the fallow in a multiplot or multi-agroecosystem facilitative arrangement.


## OVERLAPPING PATTERNS (T)

Overlapping patterns are a rotational variant that can find use with seasonal crops or with long-term systems such as orchards, tree crops, or forestry plantations. These may address the same concerns as a rotational system or be quite elaborate and have a separate classification. What differentiates them is the lack of a clear primary species.

In agroecological terms, these patterns are all semisequential intercrops that take advantage of resource surpluses at the end and beginning of the planting cycles, when the component plants are drawing fewer resources. This strategy can encourage more efficient land use by shortening the overall growing cycle through a rotational overlap and can confer biodiversity gains.

### The Standard Design

In this catchall category are some key variations. Two variations are presented.

#### Seasonal Variations

Overlapping sequential systems have one or more longer-duration species intercropped with series of shorter-interval crops. They can work through intercropping with different-period productive species. Included are systems, such as hedgerow alley cropping, where a longer-duration perennial (the hedge species) is continually matched with one or more seasonal crops.

There are a number of temporal variations on this theme. Some of the possible sequences are portrayed here:

$$p \rightarrow Ps \rightarrow S \rightarrow Sp \rightarrow P \tag{5.1}$$

$$p \rightarrow Pf \rightarrow Fp \rightarrow FP \rightarrow Fp \tag{5.2}$$

$$pf \rightarrow Pf \rightarrow Fp \rightarrow FP \rightarrow Fp \tag{5.3}$$

The first variation (5.1) uses a primary species ($p$) planted alone. As this grows ($P$), a secondary species ($s$) is added. The primary species is harvested, and at a later stage the primary species is reintroduced. The secondary species is subsequently harvested. The sequence may continue from this point. One or all of the species may be present as rootstock, but as they do not draw resources, this qualifies as a rest in the cycle. In temperate zones, this may include an overwintered crop.

The second variation (5.2) involves planting facilitative species ($f$) after the primary crop ($p$) is planted and established ($P$). From that point on, the facilitative species remains, while a sequence of primary species are planted and harvested. The facilitative species can be a perennial cover crop or a hedge (as in alley cropping). The third variation (5.3) is much the same as the second (5.2) except that the facilitative hedge ($f$) is put in place when planting the primary species.

## Plantation Variations

The basic premise in overlapping plantations is shown with two time lines. The first has a normal ending-replanting sequence; the second multiple species with overlap.

$$P \rightarrow | P \rightarrow | \tag{5.4}$$

$$P \rightarrow PO \rightarrow O \rightarrow | \tag{5.5}$$

The overall timing of these systems is the same and harvest of the first planting occurs when scheduled. The difference is that species $O$ is more valuable and slower to mature than species $P$. The symbol ($|$) signifies the end of the rotation. Some examples exist, but published documentation is lacking. Pine and oak have the complementarity to accomplish this, and this succession is found in nature. Another is Brazil nut (*Bertholletia* spp.), which requires up to 25 years before nut production commences. To maintain economic viability during this period, sequences containing shorter-duration fruit trees can help fill the income gap.

The simplest way to accomplish this is to plant the second rotation while the first is still in place. When the first is removed, time is saved as the second planting has had more time to mature.

## Notable Variations

A slightly more complex, single-species plantation succession occurs when the second rotation is planted among the first, allowed to grow to a set height, and pruned to ground level. The purpose of this cutting is to avoid damage to the new plantings when the mature plantation is removed. The existing rootstock accelerates stem growth for the new planting, saving time in reestablishment.

Some of the combinations can be very species rich. For example, in Fiji the sequence starts with yam, yaqona, and taro and, one month later, bele is planted, followed immediately by banana, pawpaw, and sugarcane. After nine months, a harvest sequence starts, first with yams and bele, followed by the longer-lasting species. The entire sequence takes up to five years (Siwatibau, 1984).

## The Design Package

These alternatives lie somewhere between a fully rotational and a simultaneous intercrop and can have, depending on the problem addressed, a distinct design package. One difference is in having the knowledge to implement a more farsighted and complex cropping sequence. The climate (e.g., yearly rainfall patterns) should allow for the seasonal variation. The longer-term systems are less climate dependent.

## Landscape Considerations

This category of intercrops offers greater biodiversity potential with lower space requirements than a simpler rotational sequence. The advantages are gains in economic and land use efficiency, a reduction in the dangerous bare land phase, and a greater ability of the landscape to combat natural stresses.

## TAUNGYAS (T)

Taungyas exist where agricultural crops (forage and grazing included) in various temporal sequences and for portions of full rotation are established and grown under tree plantations or orchards. With the amount of possible variation, these constitute a large category of temporal systems.

As with overlapping rotations, the key aim is to use surplus essential resources at various stages in a forestry or tree crop plantation to support other agronomic activities. In contrast to overlapping systems, there is one governing species around which the temporal sequence is assembled.

### The Standard Design

There are four standard variations with some latitude to combine attributes. Figure 5.1 shows the four taungya forms.

#### Simple

The simple taungya proceeds from a crop-tree mix to a monocultural perennial plantation. The end point can be a tree crop or forestry plantation. The trees are the primary species. As with any taungya, the added crops do not always compete against the primary species and can be an ecologically and economically beneficial force.

The designs can be defined through the temporal sequence of agrotechnologies. In the basic form, this is

$$ct \rightarrow T \rightarrow | \qquad (5.6)$$

where crop $c$ (the taungya species) is planted with a tree or tree crop ($t$ in the establishment phase, $T$ as a mature plant). Once the crop is harvested, what remains is commonly a single-species plantation. This is the version shown in the first column of Figure 5.1.

The initial stage can also be an intercrop, and/or the final stage can be some type of multispecies plantation. The sequence for a more complex beginning taungya form is

$$ect \rightarrow cT \rightarrow T \rightarrow | \qquad (5.7)$$

where $e$ and $c$ are the taungya species and $t$ the tree. In time, the $ec$ intercrop is reduced to a presumably more shade-tolerant single understory species ($c$) and then to a pure plantation of species $T$.

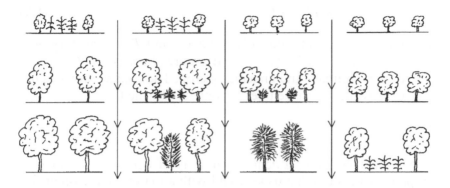

FIGURE 5.1. The four taungya subclassifications, from left to right, are (1) simple, (2) extended, (3) multistage, and (4) end stage. Key points along the temporal progression are illustrated in each column.

There is an all-forestry example that follows this pattern. It uses fast-growing, usually nitrogen-fixing species as guide species for a slower-growing, higher-value species. The closely planted guide component ensures a straight, clear, high-value stem, while providing pole crop and facilitative services.

*Extended*

In extended systems, some agricultural activity always accompanies the primary crop. This can be a tree crop or forestry species.

The first stage is always a tree-crop intercrop. This is followed by any number of nontemporal agrotechnologies, e.g., light shade with tree row alley cropping. The final stage can be a heavy shade with a shade-resistant understory. This sequence is

$$ct \rightarrow eT \rightarrow fT \rightarrow | \qquad\qquad (5.8)$$

where crop $c$ and tree $T$ (and $t$) are in the initial planting. Crop $c$ is followed by crop $e$ and then crop $f$ until the sequence ends. Any number of crop species can be sequenced. Commonly, the final understory ($f$) is a shade-resistant pasture.

As complexity increases, so does the number of options. One variation found in Brazil uses a more indirect sequence to support a perennial understory crop requiring shade trees. This sequence is

$$eb \to btc \to TC \to | \qquad (5.9)$$

where $e$ is a seasonal crop, $b$ is banana, $t$ and $T$ an overstory non-productive shade tree in early ($t$) and mature ($T$) phase, and $C$ (and $c$) is cocoa, the primary and understory crop. The banana provides initial shade while the shade trees become established.

## Multistage

A multistage taungya denotes an array of complex, overlapping cropping sequences that usually end with a mature tree crop or forestry plantation. In essence, these duplicate the complexity of natural successions and is a bio-rich version of the sequential overlap with added environmental and socioeconomic properties. Because of the long-term planning involved, these are rare.

The simplest variation is constructed around overlapping monocultural tree or tree crop plantations. Additional species are included to take advantage of plant-plant complementarity for specific essential resources that are in surplus. Schematically presented, an example is

$$ct_1 \to T_1 t_2 \to T_2 e \to T_2 \to | \qquad (5.10)$$

where $c$ and $e$ are different understory crops, $T_1$ and $T_2$ are overlapping plantation tree species, and $t_1$ and $t_2$ denote the early establishment phase for species $T_1$ and $T_2$.

## End Stage

In end-stage taungyas, as a forestry plantation is thinned, the internal competitive situation is lessened. This opens an opportunity for additional agricultural activity.

The standard design includes grazing below a thinned canopy. Seasonal cropping in the light shade environment is also possible.

$$t \to eT \to | \qquad (5.11)$$

This process starts with a single tree or tree crop species and, at a future thinning, a crop species ($e$) is added to utilize surplus essential resources, especially accumulated nutrients.

## Notable Variations

There is no reason that simple and end-stage taungyas cannot be combined where the midphase is a monocultural or multispecies plantation.

$$tc \rightarrow T \rightarrow eT \rightarrow | \qquad (5.12)$$

Another option is to maintain the simple taungya, but to use a later-stage thinning as the starting point for a new plantation sequence.

$$t_1c \rightarrow T_1 \rightarrow T_1t_2 \rightarrow T_2 \rightarrow | \qquad (5.13)$$

This combines a taungya with an overlapping sequential plantation sequence ($T_1$ and $T_2$).

## The Design Package

Despite the economic advantages, a number of conditions must exist before a taungya becomes viable. The facilities must exist for farming by the land user or conditions must be attractive enough to entice an outside farmer. In the latter case, a stringent set of land use prerequisites must be agreed upon.

Trees are the primary crop, and their value should exceed, or be equal to, that of secondary species. As such, the competitive pressure favors the trees.

If animals are used (in pastures or as a weed control), they should not eat the trees, nor should they be large enough to step on young plants. For larger trees, fencing can be in place, barriers to movement can exist, the tree can be protected (e.g., with repellents or piled brush), or the animals can be carefully watched.

End-stage taungyas are associated almost exclusively with high-value forestry plantations or where the option exists to raise larger-diameter, higher-value, and better-quality logs in combination with a lower-worth species in a multispecies plantation. Other options include pastures or crops raised in the resulting environment.

## Landscape Considerations

Seldom do long-term forestry plantations occupy high-quality farmland but, with the different taungya options, the impetus is in this direction. This is because the high level of outputs (the tree and crop) makes these highly revenue oriented. Because of their longer-term nature, taungyas require a larger area than shorter-term rotations, especially if crops are to be harvested on a regular basis.

Because of this long-term nature, taungyas can require very large areas to successfully implement. The land can be in one large holding or cross boundaries, with many participants and the need for agreement. Despite the possibilities, no examples of tightly coordinated taungyas have been found where the cropping (taungya) phase is shared across holdings.

## BIOMASS BANKS (A)

Areas specifically devoted to leafy growth can be placed to provide plant biomass for various farm uses, including animal feed (carried to corralled animals), some forms of aquaforestry, insect control purposes, and/or green manure. What differentiates these from forage or feed systems is that they are not designed as pastures or to be grazed directly, although visually they may be comparable. Because of dual use, biomass banks might have a dual classification.

Banks can be classified as auxiliary or principal-mode systems depending on the use of the biomass. If it is directed to other systems, this is an auxiliary agrotechnology (as categorized here). If used directly as a product, e.g., animal feed for sale, this is a principal-mode system.

### The Standard Design

There are two standard designs; one for mechanization, another for hand labor. The first is used with nonwoody annual or perennial plants. For cost-control purposes, these are mostly perennial systems. A second option is mini-hedge designs using woody perennials, where low hedges are conducive to machine cutting. In the hand-cut version, the plants are taller, usually about 0.5 m in height such that cutting with a machete or similar implement is easier. With either type, the biomass is transported to other locations for use.

Within this context, these systems usually use fast-growing, high-biomass producing species. Other DPCs for green fertilizer are rapid biomass decay and nutrient content in the leaves that matches crop requirements. Nitrogen fixation is often a dictated DPC. Other species, e.g., the African tree species *Tithonia diversifolia* (Buresh, 1999) with a high phosphorus content in the leaves, can serve well where this nutrient is lacking.

### Notable Variations

Where machine harvesting is employed, grass species may be a better option than pruned trees or shrubs. They are used in temperate zones where machines harvest hay for later use. The advantage of trees or shrubs over grass lies with the drought resistance of a suitably chosen tree species. They may be the better option in warmer, drier climates.

### The Design Package

The two versions have different uses and design packages. Hand labor systems are found in regions with an ample supply of labor, with sufficient land area, and where staple crops cannot be supplied with nutrients through internal agroecosystem design alone (Kormawa et al., 1995). For mechanized systems, the use may be more limited. Organic farming may be one such application.

Green biomass integrated into the soil structure may help retain moisture, thereby reducing drought risk. Other applications use appropriate agrotechnologies (such as absorption zones).

In the auxiliary role, these systems have strong locational needs. As a green fertilizer or mulch, the volume of biomass needed requires road transport or the system needs to surround or be inside the recipient plot. This is also true where biomass may be used in weed control. As an insect repellent source, the area used can be smaller and the placement more flexible.

## CAJETES (A)

*Cajetes* are designed for a specific application and consist of a line of deep, separated holes located uphill above a crop plot. The objective is to capture rainfall and make it available to the adjacent crop area through belowground infiltration. The difference between these and standard infiltration structures is the use of holes rather than long ditches and the depth of the structures (*cajetes* are deeper than infiltration ditches).

### The Standard Design

As *cajetes* are not widely used, only a rudimentary standard design has evolved. The depth and spacing of the trench is a function of rainfall and runoff. The rule is that the ditch is filled, but not overflowing, after a heavy rain.

### Notable Variations

One variation uses trees or other vegetation to line the ditch structures to prevent collapse or to slow filling with waterborne soil. Hedges planted above the ditches serve a similar function.

### The Design Package

*Cajetes* are used in intensive agriculture regions where a land shortage forces the use of marginal lands (i.e., moderate hillsides). The site situation is where brief periods of high rainfall are more a problem than a blessing, and measures must be taken to rectify and steady moisture availability.

## CATCHMENTS (A)

Areas are often put side for the sole purpose of promoting water runoff. As the water is used for purposes outside the immediate system, these are facilitative landscape additions. Often without productive intention, they can be the largest, in terms of area employed, of the auxiliary structures.

### The Standard Design

For runoff systems, two basic types exist, although in-use design standards have not evolved for either. The presence or absence of vegetation does not detract from the design purpose.

*Quick runoff*—These are areas devoid of or containing little vegetation where, because of high evaporation and transpiration rates, less water is lost when it is quickly removed. They are limited to arid zones with sparse vegetation. Quick runoff systems require water storage facilities or rapid permeation into the soil strata.

*Delayed runoff*—This is used where rainfall exceeds evapotranspiration rates and the water is used for other purposes. In situ plant biomass is the mechanism used to capture, hold, and release water at a constant rate. Natural forests are the norm with these systems (see Water Management, Chapter 6).

### Notable Variations

There are few variants with quick runoff systems in arid or semi-arid regions. In contrast, in high-rainfall areas with substantial productive potential, there are different options. Some principal-mode systems can double as a catchment. Agroforests and properly formulated heavy shade systems are among the possibilities.

### The Design Package

The reason for any catchment is water management. This is detailed in Chapter 6, but, briefly, it is best in a situation where additional water, other than natural rainfall, is needed. Outside the discussion in Chapter 6, catchment water may also be put to nonagricultural or nonforestry use.

## INFILTRATION BARRIERS (A)

Barriers can be placed in or between fields for the express purpose of slowing water and promoting infiltration. They need not have and often have no associated vegetation. Although they serve some of the same ecological functions as strip cropping, as an auxiliary technology they have no productive intent.

Barriers can be between individual plots or landscape-wide. In parts of the African Sahel, widely spaced (at 10 to 20 m intervals) ditches on almost flat or slightly sloping lands serve to contain the brief periods of high rainfall.

### The Standard Design

The standard design can be a

1. single hedge line,
2. hedge with ditch,
3. ditch alone,
4. ditch with bund,
5. ditch with bund and hedge,
6. a bund only,
7. a bund or ditch with interspaced trees,
8. a continuous stone barrier, or
9. some other single-line structure.

Commonly, barriers are parallel with the contours of the land (see Photo 5.1), but these structures can be perpendicular to the contours,

PHOTO 5.1. A contour infiltration ditch with supporting vegetation separating two wheat fields. In this case, most of the accompanying vegetation is uphill from the ditch. The amount of water captured shows the effectiveness of this structure.

crossing very shallow depressions to impede water flow and prevent soil loss.

## Notable Variations

Barriers can be allied with vegetation to encourage infiltration and provide some alternative output. Accompanying vegetation can also offer a range of facilitative services. Barriers are incorporated into alley cropping and, in modified form, in constructed terraces.

## The Design Package

Barriers have very wide applications and can exist wherever topography, rainfall, cropping systems, and soil types create conditions where water erosion can exist and/or infiltration can benefit crops.

## FIREBREAKS (A)

Firebreaks are strips of vegetation or, more commonly, bare land that impedes the spread of fire. They are often associated with forestry, but can be found in agricultural environments.

## The Standard Design

The standard design uses 2 to 3 m strips devoid of vegetation between fire-susceptible plots or between a road, railroad, or areas with fire danger and fire-susceptible tracts.

## Notable Variations

Where fire danger is low and/or fires less intense, grazing or mowing can substitute for a more formal firebreak. If a fire-resistant species is available, firebreaks can double as a living fence for plantation grazing or protecting fire-prone grasslands. Examples are species of cactus.

## The Design Package

Firebreaks are more common in regions with long dry seasons and fire-susceptible grasslands, crops, or tree plantations. As their use can be expensive, utilization is commensurate with the value of trees or crops and the degree of protection deemed necessary. The use of fire-breaks within the landscape is discussed in Chapter 7.

## LIVING FENCES (A)

Living fencing is a group of auxiliary designs that uses living plants to demarcate plots or discourage passage by large animals or people (see Figure 5.2), in contrast to dead fencing with posts made of metal and/or wood.

### The Standard Design

A number of standard designs exist depending on need and use.

> *Hedge designs*—These utilize closely spaced tree or shrub species. Among the options are plants with spines or thorns to discourage passage and/or with interwoven branches, undertaken while the plants are young, to form an impenetrable barrier. Among the desirable plant characteristics are a high degree of impenetrability along with drought resistance, grazing potential, rapid growth, and ability to produce some useful alternate product (Ayuk, 1997).
>
> *Post designs*—With these designs, living trees replace posts and fencing wire is strung on these trees. The trees can be widely spaced where branches, pruned from the trees, serve as stringers, preventing wire from sagging. Screws can be used to anchor the wire to trees so it will not become ingrown.

### Notable Variations

Among the variations are to use the living fence as a productive entity with outputs such as nuts, fruits, forage, or firewood. The post design is highly suitable as an alternate forage source, as canopies are above the reach of most grazing animals. When needed, the forage can be pollarded from the trees.

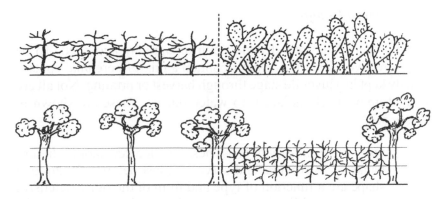

FIGURE 5.2. Different types of natural fencing. The upper illustration shows those composed of vegetation with interwoven shrubs (left) and naturally un-yielding plants (right). The lower drawing has two wire-supporting options: close-spaced pollarded trees (left); wider-spaced trees with branches interwoven into the wire (right).

## The Design Package

The living fence is less costly to install than dead fence and, for some versions, annual maintenance can be lower. Other versions are less cost effective and, in general, they may require a nongrazing period or plant protection to allow for establishment. Among the live hedge-use parameters in Burkina Faso are larger plot size, whether fertilizers and irrigation are used, cash crops, and presence of dead fencing (Ayuk, 1997).

## RIPARIAN DEFENSES (A)

Riparian buffers are a variable group of auxiliary structures designed to prevent soil and nutrients from reaching active watercourses. They function by slowing surface water and promoting the active capture of nutrients through plant absorption. Riparian buffers are most effective when they are part of a fully coordinated landscape with other water defenses and properly positioned and designed agrotechnologies.

## The Standard Design

A number of designs depend upon topography and need. To be fully effective, all buffer designs contain a mix of perennial species usually kept in a juvenile stage through harvest or pruning. Not all are located immediately adjacent to watercourses. These are shown in Figure 5.3.

> *Simple streamside buffers*—These are located along stream-banks or riverbanks and serve as a last defense. Width and shape are a function of type and form of the other defensive structures, while the vegetation is usually mixed perennial species.
>
> *Fingered buffers*—These are upslope extensions of simple buffers that occupy wadis or other areas of intermittent water flow. As with other forms, these are composed of mixed perennial species.
>
> *Arm-and-hand buffers*—These larger buffers are slightly more complex versions of the fingered buffer. As the name implies, they are connected to a simple buffer by the arm (located in large wadis) where fingers extend into wadi extensions.
>
> *Detached buffers*—Not all fingered buffers are directly connected to a simple buffer. For any number of reasons (e.g., a road or interplot access), a break in a fingered buffer may occur. This can reduce the effectiveness of a buffer, but with modification, it can still accommodate the specific need.

## Notable Variations

Buffers can also serve as a living fence, windbreak, forage or wood source, or other productive purpose. Specific-use designs also exist, where the vegetative composition can be formulated to remove a particular combination of nutrients (Wojtkowski, 2002, p. 234).

## The Design Package

Given the wide variety of options, riparian systems should exist in all landscapes. Because many of the variations employ cropping sys-

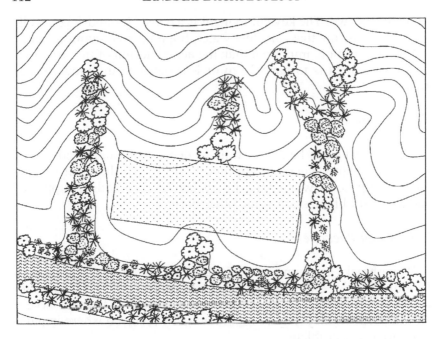

FIGURE 5.3. Four riparian variations. Along each side of the stream (bottom) are the streamside buffers; extending uphill in the wadi (left) is a fingered buffer; in the center on each side of the plot (top and bottom), is a detached or interrupted buffer; and on the right is an arm-and-hand design.

tems, they need not be outside of normal productive activities, only requiring a reordering of existing or proposed systems.

## WATER BREAKS (A)

Water breaks are strips of permanent vegetation in floodplains that transect crop plots. They are oriented perpendicularly to water movement where, during times of flooding, the purpose is to prevent soil loss and to capture and deposit waterborne soil.

### The Standard Design

Water breaks are usually strips of vegetation, either short well-rooted hedgerows, strips of taller mixed tree species, or some combination of the two. The key component is a dense, ground-level, well-anchored structure that can slow water, trapping and retaining soil.

Location is more important than the actual design of the structure. The key element is to have 20 m or less between individual water breaks.

### Notable Variations

Water breaks can be stone walls or dense well-supported fences with or without vegetation. If properly designed, they can serve a double purpose as windbreak, insect barrier, live hedge, supply of cut-and-carry biomass, or some other purpose.

### The Design Package

Water breaks are found solely on flat bottomland that is subject to periodic flooding and scouring, where the occasional flood causes severe erosion and negatively influences production. If they serve a double purpose, they may be more favorably received where inundations are less a problem.

## WATER CHANNELS (A)

Trenches can be designed to convey water from one place to another. They can employ vegetation as part of the design, either to support the trench structure or to shade the water.

### The Standard Design

Water channels divide into two main types, narrow and broad. Generally, narrow channels are found on small farms and contour hillsides. The standard design uses trees or shrubs planted on the upslope side of the trench with vines or grasses on the lower side (see Photo 5.2). This is to facilitate cleaning. A second option is to have nonwoody perennials on both sides of the trench.

The broad type is located more often on irrigated flatland. Because they have more of a landscape presence, the design options are presented in the next chapter (see Figures 6.2 and 6.3).

PHOTO 5.2. A narrow, active water channel showing uphill vegetative placement. This photo was taken in the winter before the leaves reestablished.

## Notable Variations

There is no reason why a hillside channel cannot be part of a *cajete*
system that feeds water, through infiltration, to fields below the water
channel. At the definitional fringe of this agrotechnology is the use of
vegetation to shade lined water channels or water pipes. The purpose
is to keep the water cool and, for open channels, to retard evaporation.

## The Design Package

Channels are found in areas where there is a long dry season and an
amassed source of water, commonly rivers, streams, or catchments

with or without water-holding structures (ponds, etc.). They are also found where drought risk is a factor and is countered with a water source that is available during periods of drought.

## WINDBREAKS (A)

Wind structures are a common feature in many landscapes. The purpose is to protect crops from wind damage, reduce transpiration and crop water loss, and to protect animals from temperature extremes and climatic exposure.

### The Standard Design

Two standard designs exist for windbreaks: single-species and multispecies versions. A brief description of each is given here; more detail is provided in Chapter 7.

> *Single species*—This is a row one tree wide of a single species, where the design intent is to provide windbreak benefits, but using a minimum amount of land area. Commonly used plants are nonspreading perennials that are straight and tall, with a dense lower canopy and a narrow, more open, upper canopy.
> *Multiple species*—These windbreak types employ different species to accomplish the design task. They are generally used in conjunction with larger cropped areas or where an environmental need requires more biodiversity.

### Notable Variations

Any number of systems can serve as windbreaks; therefore the number of variations is quite large. As windbreak size and density within the landscape increase, they fall under the heading of a shelterbelt (or timber belt). Both these topics are discussed in greater detail in Chapter 7.

### The Design Package

The design package is rather simple. Windbreaks find use whenever sustained or brief high winds curtail production. They may be needed for crops, fauna, or newly planted tree crop or forestry plantations. Chapter 7 details the use of these options.

# Chapter 6

# Water Management

Water is managed landscape-wide, to prevent water erosion or to prolong a cropping season through more soil moisture and extended temporal availability. Other objectives are to supply fields with irrigation water and households with drinking water, or to keep streams and rivers free of waterborne impurities. In the better-designed agro-ecological landscapes, all these objectives can be met.

Water management has a number of facets, many of which rely upon the infiltration of water into soil structure, eliminating surface runoff and associated erosion. Infiltration slows water movement and can extend the growing season or mitigate periods of low rainfall. Exposure of water to active vegetation is a means to remove water-carried nutrients. Slowing this flow increases the capture of these nutrients.

The methods used to reduce erosion and promote infiltration also moderate the effects of seasonal short-term drought. Interestingly, many of these same measures help counter and maintain the status quo during periods of extremely high rainfall. Proper design is not a plot approach, but is best as a coordinated intersystem plan.

This chapter discusses the use of agrotechnologies, both principal-mode and auxiliary, in the overall landscape, the different landscape layout options, and some unique and not-so-unique problems encountered. The chapter closes with a short case study of a water-intensive landscape from Japan.

## *PRINCIPLES*

Managing terrestrial surface water flow is accomplished through two basic mechanisms: cover and barrier approaches. These are briefly outlined here.

Cover systems function by keeping the soil continually covered and protected, with vegetative residues (e.g., leaf litter or mulch) or live plants (e.g., ground-level cover crops). Live plants have the ability to capture fertilizer runoff and eliminate both soil and nutrient pollution. The effectiveness of live vegetation as a control measure is reduced in relation to canopy height (the higher a canopy is above the ground, the less the protection), unless these canopies produce and retain sufficient ground litter or there is a large amount of surface roots.

Barrier systems are generally used where the soil must be unprotected for a period of time. Barriers can be within the soil (e.g., a ditch) or aboveground (e.g., a hedgerow). These are permanent structures that, in most cases, are continuous and run parallel with the elevation contours of a slope. Others may cross contours and/or be discontinuous. Unvegetated structures have less of a role in nutrient capture, but can still reduce soil loss and increase infiltration.

Vegetation-covered infiltration structures (also permanent soil cover) may have a less recognized utility, that of harboring earthworms and other valuable microfauna. Their value, in both soil improvement and water permeation, is well known, and covered barriers can serve as a reservoir, allowing these organisms to spread to adjoining crop areas.

## AUXILIARY AGROTECHNOLOGIES

Water management is a common problem requiring a range of solutions, and a number of dedicated auxiliary agrotechnologies have been formulated. The auxiliary agrotechnologies listed first are specifically designed for water management; others can have water management as a secondary purpose. A basic description, with variations as appropriate, of each is found in Chapter 4. The water management auxiliary agrotechnologies are

- *cajetes,*
- catchments,
- infiltration barriers,
- riparian defenses,
- water breaks, and
- water channels.

Other auxiliary agrotechnologies, those with possible secondary water management purposes, are

- biomass banks,
- firebreaks,
- living fences, and
- windbreaks.

## *Water Management Agrotechnologies*

With the exception of water breaks, water management systems are mostly barrier systems that follow land contours. Within each category, there are also topographic exceptions, e.g., some are perpendicular to the slope contours or provide cover rather than barrier protection. A brief summary of each agrotechnology here, supplements the descriptions given in Chapter 5. How these are used and placed is explained throughout this chapter.

### Cajetes

These are a series of deep holes specifically designed to hold water from brief, intense periods of rainfall and to release it into subsoil strata. These structures are usually located higher on hillsides to supply water to crops in an adjacent field at a lower elevation.

### *Catchments*

These serve to capture water for use elsewhere. In arid or semiarid regions, they may be devoid of vegetation or have only enough associated plant life to filter impurities from the water.

In wetter regions, vegetation is meant to hold sporadic high rainfall and release it at a constant rate. Naturally occurring forest ecosystems can accomplish this task, as can specific agrotechnologies. On a larger scale, catchments become watersheds. These are discussed in a subsequent section.

### *Riparian Defenses*

The most basic of dedicated structures are riparian defenses. As shown in Figure 5.3, a number of different designs exist.

Generally, by being located along streambanks, riparian buffers occupy the lowest elevations of the landscape, serving as a last defense again stream contaminants (soil runoff included). These are usually composed of mixed perennial species that are pruned to stimulate growth and nutrient uptake.

Extensions of the riparian buffers follow normally dry wadis. These are fingered buffers and, at higher elevations where dry wadis converge, the area may be occupied by permanent vegetation. These perched buffers can serve other landscape functions, including nutrient infiltration and as absorption zones.

## Infiltration Barriers

Single-row barrier structures are placed to slow the movement of water and promote water infiltration into deeper soil strata. They can take many forms, from ditches to bunds, from grass to hedgerows or any combination of these. Normally, these follow land contours (see Photo 5.1), but a design variation has them crossing, at right angles, shallow and usually dry watercourses.

Figure 6.1 shows some different placements of barriers. The types are contour, counter- or cross-contour, fingered, and two kinds of semicontour placements. The fingered barrier is also a riparian system, but used in situations with very moderate slopes.

These variations can address different needs or can be merely an alternative to accomplish the same task. On steeper sites, the better options are contour and semicontour barriers, with the fingered barriers used as a final defense to further extract nutrients and contain soil movement.

For shallow locations, all are applicable and all can double as progressive terraces. For example, the counter-contour design (Figure 6.1), observable in Mexican landscapes, uses low rock walls to cross shallow valleys, serving as a modest form of a progressive terrace.

## Water Breaks

The need to control floods, through water breaks, in a valley is a different objective. They can have double purposes as a windbreak or a living fence in addition to their primary role in holding the soil on a site and capturing dirt being washed from upstream locations. Water break use in Mexico has been described (Nabham and Sheridan,

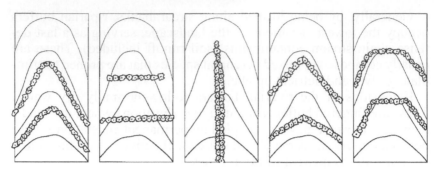

FIGURE 6.1. The placements of infiltration barriers with regard to slope contours (with the highest elevation being at the top portion of each figure). From left to right, the placements are contoured, cross-contoured, and fingered (riparian), with the two drawings on the right showing different versions of a semicontour placement.

1977), where willows woven in place provide this double use in bottomland locales subject to flooding.

## Water Channels

As part of water management, channels carry water from one part of a landscape to another. The water can be for noncropping purposes (e.g., household and animals) or for irrigation use (mostly flood irrigation), and originate at a catchment or from another source. Storage ponds may be part of a regional or farm hydraulic system. Mostly these are gravity based, relying on a contour placement to achieve their purpose.

Other forms of irrigation (sprinkler, drip, etc.) normally do not employ water channels. Disbursement through piping is generally a plot rather than an overall landscape influence.

*Design Options.* As a landscape structure, water channels do not take much space. The various design criteria, especially with vegetation, are briefly described in Chapter 5 and expanded here. As mentioned, there are two types, narrow hillside and broad flatland.

Narrow, less obtrusive structures (see Photo 5.2) are generally found on small farms in hilly locations, more frequently with subsistence agriculture. Nutrient-rich sludge, along with water, can be provided from the channel to adjacent crops.

Another aspect of narrow channels is vegetative placement. Cover crops and riparian buffers are common in this context. Closely spaced trees and shrubs are usually located on the uphill side of a channel. This is done for bank stability and to facilitate cleaning. For channels that parallel steep slopes, trees or shrubs may help support the lower dike, but with a wider spacing. Again, this is to simplify cleaning.

Since some infiltration is associated with earthen channels, they can be incorporated into a productive landscape design, where trees or shrubs are located below and adjacent to this water source (as with *cajetes*). Another use is as a buffer structure where, e.g., windbreak trees that are not overly complementary with neighboring crops are separated from them by a water channel. This is especially useful when water is the limiting resource in the tree-crop pairing.

Broad channels, usually on flat sites, through the design options, have more of an agricultural presence. Among the design variables are embankment form and vegetative placement. Figure 6.2 shows five bank types:

1. Natural sloping
2. Precipitous
3. Broad dry
4. Broad wet terraced
5. Narrow terraced

The first of these, natural sloping, is the standard and the foundation for riparian placement. The second, precipitous, is used where land is at a premium and space, especially high-fertility sites, cannot be wasted. The actual bank can be earthen, stone, or tree supported, as with some terrace structures.

The other types, broad dry, broad wet, and narrow terraced, support subsidiary agriculture. The broad terrace often serves a more commercial application, taking advantage of a moisture-rich, potentially prolific site for moisture-loving crops. It can be dry, with close subsurface moisture, or a paddy, with surface water. These terraces may find use in arid regions where channel locations allow year-round cropping.

Another design topic connected with broad water channels is embankment erosion protection. As part of river and stream management, it is presented later in this chapter under Landscape Layouts.

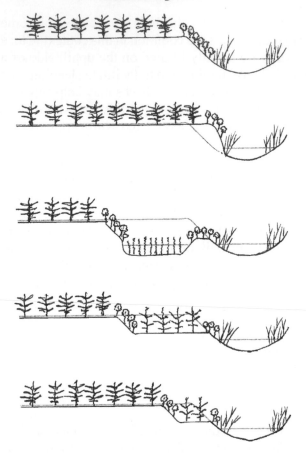

FIGURE 6.2. The five water channel bank designs, from top to bottom, are (1) normal with riparian vegetation, (2) precipitous, (3) broad terraced—paddy, (4) broad terraced—dry, and (5) narrow dry terraced. (*Source:* Modified from Melman and Van Strien, 1993.)

*As a Landscape Influence.* How water is allocated within an irrigation system influences field layout, content, and channel placement. The easiest structure to design is entirely within one holding, managed by one land user.

Most larger systems span different holdings and require the cooperation of a number of land users, often with the need to equitably divide the resource. In addition to the need for social organization

(Trawick, 2001), there is also the need to consider the placement of the channels, location of fields, and the types of crops planted.

With multiple participants, one problem is in dividing the resource. This is less an obstacle where measuring devices are in use, e.g., a crude paddle wheel with a revolution counter may suffice. In more traditional societies, other means have been devised. Some channel structures physically divide this resource; others utilize temporal sharing (Geertz, 1972).

Some clearly identifiable distribution patterns are found. Presented here are (1) linear, (2) branching, and (3) progressive fill (see Figure 6.3). There are pluses and minuses associated with each allocation system.

In the linear channel system, the fields of the various land users are located along the flow structure. Depending on water volume, each is allowed to irrigate a certain number of plots or have exclusive access to the water for a set time period. From a layout perspective, this is the most flexible for distribution, but less so in providing for each land user's optimal needs (i.e., where a plot next to the channel might not need the water, but with few options to allocate the water elsewhere).

With a branching structure, each land user oversees a set of trenches off a community channel and a distribution node. Because the node (Figure 6.3) can be designed to ensure a more or less equal flow down the different trenches (given outlets of equal dimension and elevation), being more evenhanded, it is conducive to flow, rather than temporal, allocation. Another advantage is that each land user has more choice as to which plots actually receive the water.

The third option, the fill system, is more favorable to, and finds use in, aqua-agriculture (e.g., rice paddies), where a water level must be maintained. The spillover automatically fills a lower plot. As an allocation system, the basis of distribution is gauged by area or plots filled. These can be found on hillside locations and may be best where land users have adjacent plots at different elevations.

It should be noted that the option exists to combine these approaches. For example, a temporal distribution in a branch system may then go to an individually owned fill system. If insufficient water exists to completely saturate all the plots, less water-intensive crops may be planted on the lower elevations such that these are watered through overflow or the in-soil movement of percolated moisture. Although less studied, how allocation occurs, from both the social and

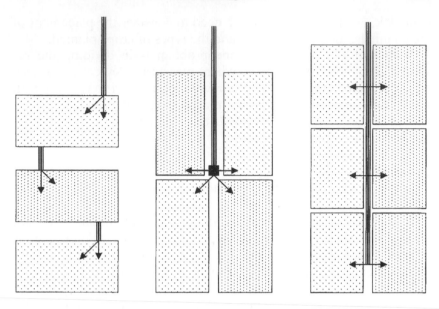

FIGURE 6.3. The three different irrigation layouts, from left to right, are (1) linear, (2) branching, and (3) progressive fill.

physical perspective, can have a profound effect on the overall landscape.

### *Other Auxiliary Agrotechnologies*

Most of the identified non–water management agrotechnologies (i.e., biomass banks, firebreaks, living fences, and windbreaks) can have some water management function. Given that auxiliary systems have little productive function, dual-purpose use is always advisable and, if they can be redesigned to contribute to water management, all the better.

## PRINCIPAL-MODE PROPERTIES

Dedicated auxiliary water management systems have advantages over principal-mode systems in that, shorn of a productive role, there is more flexibility for a more purposeful design. However, with the margins of a productive system, there can be considerable design flexibility to provide effective water management services. The advantage is that special structures may not be needed and the full productive potential of the landscape can be utilized.

Table 6.1 shows different principal-mode agrotechnologies. Those marked W are principal-mode systems formulated to cope with specific water stresses.

TABLE 6.1. Some water management properties for the principal-mode agrotechnologies

| Agrotechnology | Function | | |
| --- | --- | --- | --- |
| | Erosion control | Infiltration | Nutrient capture |
| Absorption zones (W) | G | G | G |
| Agroforests | E | E | E |
| Aqua-agriculture (W) | G | G | G |
| Aquaforestry (W) | G | G | G |
| Alley cropping (hedgerow) | G | G | G |
| Alley cropping (tree row) | G | G | G |
| Entomo-systems | V | V | V |
| Forage systems | G | G | G |
| Intercropping (multiseasonal) | G | G | G |
| Intercropping (seasonal) | P | P | V |
| Isolated tree | P | V | V |
| Living fences | V | V | V |
| Microcatchments (W) | G | G | G |
| Monoculture (perennial) | P | P | P |
| Monoculture (seasonal) | P | P | P |
| Parkland | P | P | P |
| Root support systems | V | V | V |
| Shade (heavy) | P | P | P |
| Shade (light) | G | G | G |
| Strip cropping | G | G | G |
| Support (perennial) | V | V | V |
| Support (temporary) | V | V | V |
| Terraces (all forms) (W) | E | E | E |

*Note:* E = excellent; G = good; P = poor; V = variable depending on the specific design used. This describes their overall capacity, including the frequency of the more susceptible bareground phase.

*Absorption zones* (W)—This agrotechnology is designed to conserve water and prolong a growth period. In this role, it provides good protection in and near the structures but, outside this zone, protection can be lacking unless a cover crop is used.

*Agroforests*—These often epitomize the best possible in all ecological categories. Well managed, they offer full protection in all water management categories. This is an excellent system for use in a vegetated catchment area.

*Aqua-agriculture* (W)—This is a water management agrotechnology designed more for retaining standing water than for direct protection. Because of location, it offers little in the way of erosion control, but can be useful to capture nutrients. It can be erosion susceptible and require catchment areas.

*Aquaforestry* (W)—As with aqua-agriculture, this design is more for standing water situations. As such, it offers little erosion control, but can capture nutrients and, at higher elevations, can store water for aqua-agricultural systems.

*Alley cropping (hedgerow)*—Well designed and maintained, alley cropping can be useful in all water management categories. In essence, these are infiltration structures where the row orientation can be contour, semicontour, and counter-contour.

*Alley cropping (tree row)*—This is less useful as erosion control, unless accompanied by pronounced barriers. It is useful to capture nutrients not used by companion crops. Although less effective as infiltration structures than a hedge design, tree rows can serve this function on relatively shallow sites.

*Entomo-systems*—These systems have few inherent water management properties but, due to a rather open-ended design, can be formulated for this and other purposes.

*Forage systems*—These can be formulated for water management. They capture nutrients well but, because of the presence of animals, serve as an upper-level, rather than streamside, riparian buffer.

*Intercropping (multiseasonal)*—The effectiveness of intercrops in water management depends on species, groundcover, and canopy height. With sufficient cover crop or ground litter, they provide excellent soil protection. Where wide niche coverage exists, they can effectively capture nutrients and can be positioned accordingly.

*Intercropping (seasonal)*—In a bare-ground phase, seasonal intercrops are susceptible to erosion. Depending on cropping sequence, they can be formulated to accomplish some minor water management tasks, but only in league with an auxiliary system.

*Isolated tree*—The susceptibility to natural stress is more a function of the crop system in which the isolated trees are located. Within the landscape, for seasonal crops, isolated trees are a flatland system where the trees are located in depressions or along the bottoms of wadis to promote drainage or serve a riparian function. With perennial crops, they have more locational flexibility.

*Living Fences*—These offer good, but limited protection. Good in that hedge or tree rows serve well, limited because these fences are generally widely spaced. As living posts supporting wire, they are nowhere near effective.

*Microcatchments (W)*— These are water management systems designed to increase the capture of water on hilltops or less steep hillsides. Given sufficient planting density, they have good water capture and erosion control properties.

*Monoculture (perennial)*—Water properties depend upon the crop used but, with some redesign to established principles (such as heavy ground litter), perennial monocultures can play an effective water management role.

*Monoculture (seasonal)*—As the most susceptible to erosion and nutrient runoff, short-term monocultures have few positive water management attributes. Even on flatlands, they require the protection of other systems to be ecologically integrated into a farm landscape.

*Parkland*—As with isolated tree systems, parklands have a role in draining standing water and protecting wadis. With seasonal crops, they are susceptible to water-related problems and, within the broader landscape, must be treated accordingly.

*Root support systems*—These have few inherent positive water management properties, while not making a situation worse. They can be a vehicle for nutrient retention in high-intensity orchards and a source of protective ground litter. These also require the good auspices of neighboring systems to be ecologically integrated into the landscape.

*Shade (heavy)*—The water management properties of shade systems depend on design, accompanying crops, ground-cover, and a host of other factors. With good groundcover, they are often robust enough to serve as a catchment system. Any weaknesses can be countered through redesign or a neighboring system.

*Shade (light)*—Because the increased light allows for greater light penetration, there is more potential for groundcover and erosion control. Nutrient leaching depends on intensity where, in the more intense forms, more nutrients applied means more nutrients lost to the surrounding environment.

*Strip cropping* (W)—This system is designed for water management with good erosion control and other properties. If poorly designed, it can be overwhelmed, if subject to high rainfall, through erosion and nutrient loss.

*Support (perennial)*—These systems depend on the cropping combination and layout for water management. They generally equal the protection provided by any perennial system. The gains may lie in nutrient runoff protection.

*Support (temporary)*—These systems can be susceptible to a range of water-induced degradation, but are generally better than a single seasonal crop. In conjunction with other agro-technologies, such as strip contours, the possibilities are expanded, as are the possibilities for placement.

*Terraces (constructed)* (W)—Since their purpose is water protection, terraces offer a range of such services that can be augmented through vegetative additions. They are used on the most susceptible sites. During construction, they are vulnerable to water-related problems unless proper countermeasures are employed.

*Terraces (progressive)* (W)—As water management agrotechnologies, progressive terraces can be used on the most severe sites. They have some susceptibility in their construction phase but, because of vegetation use, it is less than with constructed terraces.

The suggested hillside placements of the agrotechnologies are shown in Figure 6.4. Although subject to interpretation with regard to

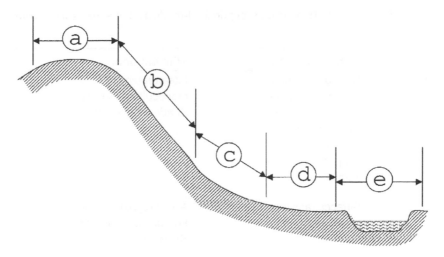

FIGURE 6.4. A hillside cross section showing different use zones. As listed in Tables 6.2 and 6.3, agrotechnologies can be paired with a use zone depending on how well they resist water-induced stress.

form, content, management, and surrounding systems, this provides a rough guide to usage (see Tables 6.2 and 6.3).

## *LANDSCAPE LAYOUTS*

The control of water runoff in a landscape is best accomplished through a series of water control structures, each placed to take full advantage of its water management properties. The idea is that, with high rainfall, each plot should be self-sufficient, being able to handle the rain that falls within an uphill area, directing little, if any, surface flow to adjoining, lower plots.

If this is not the case, an auxiliary defense may be installed as part of, or between, sections. This defense can be a contour strip, trench, or hedgerow. The type of structure used depends on the use context. A trench of sufficient depth may require mechanization, and a hedgerow requires annual pruning, while strips use more land area.

For water, the SIZs are small and downhill from a structure, becoming smaller as rainfall intensity increases. The exception is silt deposits in flow or flooding situations.

TABLE 6.2. Suggested placements for principal-mode systems (use zone shown in Figure 6.4)

| Use zone | Without problem | With caution |
|---|---|---|
| a | Absorption zones<br>Alley cropping (any type)<br>Forage (feed) systems<br>Microcatchments<br>Strip cropping<br>Support (any type)<br>Agroforests | Intercropping*<br>Monoculture (any type)*<br>Shade (any type)* |
| b | Terraces (any type)<br>Agroforests | Alley cropping (any type)<br>Forage (feed) systems<br>Shade (heavy) |
| c | Absorption zones<br>Alley cropping (any type)<br>Microcatchments<br>Support (perennial)<br>Strip cropping<br>Forage (feed) systems<br>Shade (any type)<br>Agroforests | Intercropping<br>Monoculture (any type)<br>Support (temporary) |
| d | Alley cropping (any type)<br>Forage (feed) systems<br>Intercropping<br>Microcatchments<br>Monocultures (any type)<br>Parkland<br>Shade (any type)<br>Support (any type)<br>Agroforests | |
| e | Aqua-agriculture<br>Aquaforestry<br>Agroforests | Forage (feed) systems<br>Monocultures (perennial) |

*The prospects are improved if used with an appropriate auxiliary structure (*cajetes,* infiltration ditches, etc.).

TABLE 6.3. Suggested placements for auxiliary systems (use zone shown in Figure 6.4)

| Use zone | Without problem | With caution |
|----------|-----------------|--------------|
| a | Biomass banks*<br>Catchments<br>Infiltration structures<br>Windbreaks[†] | |
| b | Biomass banks*<br>Catchments (infiltration) | Catchments (runoff)<br>Infiltration structures |
| c | Catchments (infiltration) | Biomass banks<br>Riparian (wadi version) |
| d | Biomass banks<br>*Cajetes*<br>Infiltration structures<br>Riparian (wadi version)<br>Windbreaks | |
| e | Biomass banks<br>Infiltration structures<br>Riparian (streamside)<br>Waterbreaks<br>Windbreaks | |

*To prevent drying through transpiration.
[†]As used in a secondary role, the primary purpose is cut-and-carry biomass.

## *Elevation (Contour) Placements*

In most cases, a series of appropriate agrotechnologies is placed along or across slopes. With water structures, planning and implementation initiates in the upper reaches (higher elevations). Each stage has a designated role in slowing and directing water runoff. The design corresponds with purpose and elevation.

Water management structures need not be passive (nonproductive) and, through the use of vegetation, they can also serve to manage nutrient runoff. When nutrient runoff is the primary objective of the defensive structures, they are best where biodiversity ensures that a range of nutrients are captured and when some productive output (e.g., fruit or forage) prevents nutrient buildup in the structure. Water management functions are often integrated into plot designs and, more often than not, the effectiveness of water management structures (riparian and otherwise) is a corollary of interplot effects.

## Upper Reaches

These are those flatter areas found on the tops of hills or slight slopes. As these sites are not generally at a high risk for erosion, they do not always benefit from drought control measures. As they are away from active streams, they can be overlooked in terms of water defenses or their role in a riparian system. However, this is where water control begins.

As these sites do not completely test the strength of a water defense, a number of options, including lesser systems, exist for these areas. Some combination of barrier and cover approaches is possible, but other, more productive, options can prove equally advantageous.

## Intermediate Slopes

These sites, with their steep slopes, generally provide a more severe test of a water control defense (see Photo 6.1). Where good-quality land is not intensely used, most farmers will leave these areas in permanent vegetation. The most common is natural forest, usually exploited, but not to the degree that erosion control is compromised.

In this situation, some nutrient leaching could be tolerable and even beneficial, especially when crop plots are located directly below the slope. This is more of a theoretical abstraction derived from observation, as little empirical work exists to validate any gains from nutrient leakage. If intense cultivation is required, this type of site can require a highly revenue-oriented barrier system such as terraces.

## Lower Slopes

The level ground between hillsides and along a watercourse is prized farmland and, after more spacious bottomland sites are used, can be the next area cultivated. If the stretch is comparatively roomy, farmers need not exploit steeper areas, and larger, streamside, simple riparian buffers, along with vegetated upper slopes, may accomplish all the needed water management objectives.

PHOTO 6.1. A hilltop pasture with a forestry plantation protecting the steeper intermediate slopes. As with most in-field usage, not only is this a water management layout but also the taller plantation provides wind protection to the pasture on this exposed site.

At the lower elevations, the objectives differ slightly from steeper slopes as do the structures utilized. Nutrient capture has a higher priority, while infiltration may be less a factor. This dictates the type and placement of vegetation.

If an area is less than adequate for farm needs and more marginal areas will be cultivated, there is less occasion for using simple riparian buffers. In this situation, there is more need for a range of uphill riparian buffers or similar structures that protect and thereby free more of the lower slopes for cultivation.

## Bottomland

Erosion and water control are not always associated with sloping sites. When lacking permanent groundcover, flat stretches have many of the same problems and, in addition, can have unique drainage problems. Without a slope, erosion is easier to manage, but there can be a tendency to overlook and accept land degradation on flat sites without taking appropriate measures. Water breaks are one such protection for these sites.

In addition, countermeasures similar to those used on slopes find use, e.g., hedges and bunds. Changes in tillage (timing and type) can be effective to better manage high-risk cultivation periods. An accompanying cover crop or bioresidue (tree or crop) from an earlier planting period can augment these changes.

Shallow depressions can collect standing water and, for some crops (e.g., maize), this can reduce yields. Placement of deep-rooted trees in affected zones (e.g., isolated or parkland trees) can speed absorption while providing other ecological benefits.

Any wadis that exist, no matter how shallow, should benefit from protection. If water surges are a negative factor, then the erosion countermeasures used for active streams and rivers (see the following section) may be employed. Wadis are also an ideal location for strips of natural vegetation serving numerous ecological functions.

## *Streambanks and Riverbanks*

The discussion earlier in this chapter on water channels has direct application with active streams, brooks, and rivers, especially when cultivation and an appropriate design are needed. Natural watercourses suffer from the effects of scouring and unstable wet soils and can have more need to stabilize the embankments.

There are a number of bank protection measures. With the exception of stone stabilization, most rely on the establishment of vegetation. How this is done varies (Wells, 2002). These measures can include direct vegetative establishment through cut and placed branches or stems (where cut rootstock of live woody plants are rooted in place).

Dead branches find use in various configurations, e.g., tied in bundles, as brush mats, as rows stuck in the ground parallel or perpendicular to a watercourse, or arranged in cribs at and below the water level. Logs, anchored in place at water level, also serve a purpose. Each design has an appropriate use, addressing a specific situation. Once this stage has passed, the in-place vegetation is treated and managed as a riparian buffer.

## Cross-Elevation Placements

Cross-sectional placement has focused upon those systems placed parallel to and along contour lines. Not all systems extend along land contours, and effective use can be made of noncontour placements.

Among systems oriented perpendicular to a slope are the different forms of fingered riparian buffers. In contrast with the relative design independence for contour systems, these systems must have use complementarity, and function best when mutually reinforced with contour defenses. As with all auxiliary structures, they can also have a minor productive role in the landscape.

Most options center upon riparian systems, including simple, continuous and detached fingers, and arm-and-hand designs (as shown in Figure 5.3). Their use depends upon the slope contours, the location of seasonal plots, and other landscape influences.

# SINGULAR SITUATIONS
# (AND OVERCOMING OBSTACLES)

There are circumstances where caution or revamped management is in order. These are common enough to merit some attention. More important, they provide insight into problem solving at the landscape level.

## Landscape Evolution

As land-use intensity increases, the impact of water becomes more of a concern. Figure 6.5 shows the process and how, through placement, change can be less problematic. This also works in reverse where, as populations shift to other activities, land-use change re-establishes more of the catchment function. As logic suggests, high-risk sites merit greater protection.

## Watershed Management

Watersheds require additional guidelines and, depending upon use, have more stringent needs. If water is destined for irrigation, nutrient content may not be all bad. For drinking water, purity is the norm.

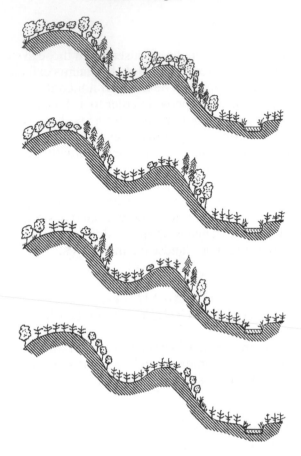

FIGURE 6.5. A cross section of hillside showing the evolution (top to bottom) from a natural ecosystem or low-intensity forestry to intensive agriculture. The placement of vegetation (as suggested through the agrotechnologies in Figure 6.4) prevents land degradation and can help preserve the catchment function.

For the water supply of New York City, intense watershed management has been found to be more effective than water treatment. This does not mean water is not treated, only that this process is less costly starting with cleaner water. This course of action is also valid in less developed regions.

Generally, watersheds in heavily vegetated regions are best if natural ecosystems are kept in an untouched state. There is some debate on this, as empirical studies do not conclusively support the advantages of forests over cleared lands (e.g., Ataroff and Rada, 2000), but anecdotal evidence clearly favors untouched lands (Sandström, 1998).

Within this context, there is some latitude to use natural processes to improve the runoff situation. Hunting may keep grazing animals from excessive destruction of natural flora. Animals such as beavers (found in Europe and North America) might be encouraged as they dam streams, decreasing erosion, promoting infiltration, and increasing the water-holding capacity of the land.

This preference for untouched land may not extend to arid regions. On nearly level arid or semiarid sites, widely spaced (at 10 to 30 m) infiltration barriers find use. These prevent water accumulation, runoff loss, and resulting erosion damage during brief periods of high rain. Examples exist in the African Sahel (Carucci, 2000).

Also, associated in-gully dams are used. To prevent washout during construction, especially if it takes some years, these dams are first constructed high on hillsides, in the smallest water channels, proceeding downward to the larger wadis. The ultimate design (spacing, types of dams or infiltration structures, and the use of vegetation) depends upon whether infiltration or surface storage is the best retention mechanism.

## Floodplain Management

In landscapes, some unique situations require special consideration. Floodplain management in bottomland areas, those that are inundated on a seasonal basis, is one of these. The annual flooding of the Nile River, with accompanying gains in soil fertility, is often given as a reason for the rise of civilization in ancient Egypt.

Where annual flooding occurs, guidelines have been proposed for better land management (modified from Nebel, 2001):

1. Habitats (e.g., dwellings) should be scattered or put on higher sites to minimize overall impact on water quality.
2. Chemical use (herbicides and insecticides) should be restricted, again to reduce runoff into water with negative effects on river flora and fauna, remembering that, in these situations, fish are often an important food source (e.g., Kvist and Nebel, 2001).

3. A permanent patch approach to landscape design (checker-board, strips, etc.) may be best, as this minimizes soil loss and maximizes nutrient accumulation (which water breaks accomplish on occasionally flooded sites). If the nutrient inflow is sufficient, this eliminates the need for slash-and-burn or swidden fallows.
4. Areas should be avoided where current and high water flow (and accompanying erosion) endanger the site.

The natural ecosystems in these areas are well adapted to perturbations and recover quickly from any change. Also, fishing as a food source should figure into any management scheme.

### Floodwater Spreading

After areas are inundated, floodwaters recede, leaving pools. They provide an immediate water source when drying starts, helping to replenish groundwater and recharge aquifers. The same effect can be realized through a planned event.

Floodwater spreading occurs when seasonal floods are encouraged through channels and other topographic modifications to cover a wide area. The purpose is to increase potential cropping without the use and cost of direct irrigation. As part of the strategy, artificial pools can retain some of the water for later agricultural use.

This strategy finds use in arid zones where, with the lack of water, little grows. The addition of flood periods will enrich the area with moisture-loving species (natural and/or agricultural) adapted to seasonal drought. The sediments left on the land are said to resist wind erosion (Kowsar, 1992).

### Salt Accumulation

The introduction of salt into crop lands is often associated with water movement (some can be windborne). This can occur when irrigation water has a low salt content and after the water evaporates or is transpired, the salt remains. This can build up and eventually, when the concentration becomes high enough, the affected plots can become unusable. This problem is rectified through flushing, where the plots are flooded and, by way of improved drainage, most of the salt is washed away. Except for the water requirement, this is less a landscape concern.

The other means by which salt may be introduced is internal to an area. In the case, the groundwater is salt laden and, as the water table rises, the salt comes into contact with crops. This is prevented by not allowing the water table to rise. Flooding through irrigation is a method of prevention. Where flooding is not possible, high transpiration, salt-tolerant surrounding vegetation can accomplish this by keeping the water table low (Schofield, 1992). These crops are watered only through rainfall or surface irrigation.

## Vegetative Establishment

One obstacle to overcome in arid zones is the problem of establishing long-term plantings. Even with species adapted to an arid climate, a successful planting is not guaranteed, and often measures must to taken to ensure a fruitful outcome.

Among the techniques to improve the success rate are the use of microcatchments, absorption (sponge) zones, and a topographic approach. The topographic approach is landscape-wide, where initial plantings occur on lower slopes and in wetter wadis. When these prove successful, then the next phase occurs further up a slope or wadi. The protection provided by the earlier plantings will hopefully have a positive influence on the new additions. The overall water management of an area may be, and is often, part of any planting scheme.

The various techniques do increase planting costs. The alternative may be yearly replanting, hoping for better-than-average rainfall distribution that enables plant survival.

## SATOYAMA LANDSCAPE: A CASE STUDY

This case study is the satoyama landscape of Japan (Attenborough et al., 2000). These landscapes are found in the region around Lake Biwa. In this example, terraces of rice are economically, ecologically, and agroecologically linked with the forested mountains and the lake ecosystem through a series of mutually supporting agrotechnologies.

The mountains provide the water catchment for rice paddies, which have, as part of area hydrology, holding ponds and water channels. A number of agrotechnologies are represented, including aqua-agriculture (with rice fields and with carp raised in the holding ponds), grass-based terraces and channels, and, with the occasional persimmon tree, a parkland system. The irrigation system is a part fill type with some connecting distribution channels.

A component of the landscape is the woodland plots that, in addition to an ecological contribution (i.e., for predator insects, water catchment, climatic modification, fauna habitat, etc.), also serve as low-intensity agroforests. In addition to some fruit trees and a supply of firewood, these plots have oak trees, which are coppiced to provide logs for raising shiitake mushrooms. The woodland plots and nearby mountains provide nesting sites for birds such as the black kite, which hunt for rice-eating rodents.

Among the abundant insect species (over 1,000), a number of predator-prey relationships exist. Spiders and other predator insects help control herbivore insect populations, which exist within grass terrace banks, along irrigation channels, in woodlands, and in holding ponds, which serve as reservoirs and travel or spread corridors. As a supplementary addition to the rice fields, catfish live in the nearby lake and breed in rice paddies. The young fish feed off and control insects (mosquitoes included). Other fauna, such as frogs and turtles, are also favored by a mix of ecosystems and provide an insect control contribution.

This is a spatially oriented landscape, one that relies little on rotations and more on the placement of agroecosystems. Much depends on the movement of the elements (water, predator fish and insects, etc.) across the landscape through an assortment of ecosystem types.

This landscape has been in active existence for thousands of years, so the system should be well evolved and in harmony with both the ecology of the region and agroecology of the productive systems employed. Given the time frame, each component, whether flora or fauna, has experienced some evolutionary pressure or has been involuntarily selected to conform to the functioning of the overall landscape.

# Chapter 7

# Wind, Frost, and Fire

The negative effects of wind on crops are well documented. With wind, humidity will decrease, plant transpiration will increase, and a site will become drier, with the possibility for yield reductions. Without wind protection, the increased transpiration has lessened yields in soybeans by 10 percent (Miller et al., 1991).

There are also some other direct consequences; strong winds can cause leaf and branch rubbing. Indirectly, windblown dust produces a sandblast effect, where sharp soil particles pierce branches and leaves (Cieugh et al., 1998). Sandblasting as well as branch and leaf rubbing has been shown to reduce wheat yields by 11 percent, maize yields by 28 percent, and bean yields by 45 percent (Brenner, 1996).

Another negative outcome can be lodging. Fallen stems expose leaves and grains to rodents, insects, and decay while increasing harvest costs. For domestic animals, wind can increase exposure, multiplying the effects of heat or cold. This results in lower weight gain and increases in mortality.

Wind is the culprit, and all those negative effects can be countered with windbreaks and/or through some form of intercropping. The latter provides facilitative wind reduction and/or anchors roots more firmly (Davis et al., 1986). By limiting seed dispersal, wind control has been thought of as a means to control weeds. Evidence points in the opposite direction, as windbreak structures can attract seed-dispersing birds (Harvey, 2000).

Related topics, frost and fire, are also covered in this chapter. Frost is a concern in temperate climates, at elevation in the tropics, and in arid and semiarid zones. This introduces another element into landscape design and a range of countermeasures.

Burning is a well-recognized land management practice with many applications, although uncontrolled fire is detrimental to the productive uses of land and an acknowledged risk. This relates to wind, as reduced air flows slow the spread of fire, and some wind control structures can serve as a partial firebreak. Other countermeasures are more specific.

## WIND EFFECTS

A number of negative wind effects need to be addressed:

1. Horizontal movement
2. Elevated gusts
3. Wind tunnel
4. Microburst and swirl

The most common forms of horizontal movement are constant breezes that rob fields of moisture, cause leaf and branch rubbing, and cause sandblasting from airborne sand. All these physically harm plants and, if forceful enough, can topple plants.

The two forms of wind tunnel are (1) horizontal and (2) vertical (see Figure 7.1). A horizontal wind tunnel occurs when winds encounter a taller obstacle and are concentrated and intensified at ground level or between canopy strata. With some forestry and agroforestry systems, this can occur when an upper canopy blocks upper air movement, concentrating and constraining the wind under the canopy and through an area occupied by crops.

The second form, the vertical wind tunnel, develops when wind accelerates down alleys as the air is concentrated as it goes parallel to tree rows. This is also associated with roads, firebreaks, or other cuts through tall plantation blocks or even windbreaks. Photo 7.1 shows a taungya system with a susceptible area countered using a nonsusceptible crop.

Side swirls are associated with wind tunnels. This is the consequence of strong wind being compacted as the air rounds the corner of a dense block of vegetation or a farm building. The wind can have such force that, in an adjoining field, crops are wind damaged or flattened. This often occurs in a clearly defined zone (see upper drawing, Figure 7.2).

Other negative effects are caused by microbursts or swirls. These are sudden vertical gusts that accompany storms and can flatten crops in fields. Often, only a small area is razed within a larger unaffected

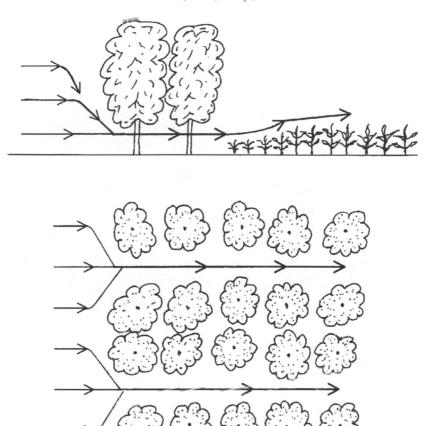

FIGURE 7.1. Two versions of the wind tunnel. The upper figure shows how wind volume and velocity increase when concentrated under tree canopies. The lower drawing shows a similar effect when wind is channeled down wide rows between tall, standing vegetation.

plot. These strong gusts are not always at ground level. These rapid bursts from storms or sudden winds generated and concentrated by flowing around obstacles can damage branches, knock down fruit, or even push over trees. A similar occurrence is top swirl where, if a vegetative block or windbreak structure is wind impenetrable, a strong wind can dip into an adjoining area.

PHOTO 7.1. An open cut caused by the presence of power lines is prone to wind tunnel air movement. The various stages of the extended taungya along the sides offer strong individual and mutual protection. Within the cut, where trees cannot be planted, a wind-resistant species, artichoke *(Cynara scolymus)*, has been planted.

## PRINCIPAL-MODE SUSCEPTIBILITY

Most agrotechnologies are wind vulnerable and are also capable of providing some form of protection to neighboring plots. For example, shade systems can protect neighboring fields from high winds but, without some protection, they may be susceptible to a wind tunnel. This is landscape mutualism promoted through pairing.

The type of crops planted also influences placement. Some seasonal crops can be adversely affected by drying or windborne particles, while others are have low transpiration rates or resist stem and leaf damage. Table 7.1 lists the relative susceptibility of agrotechnology systems to different forms of wind.

What further complicates landscape design is the relative area of each agrotechnology. Large-area agrotechnologies may confer greater protection or become more sensitive to wind stresses.

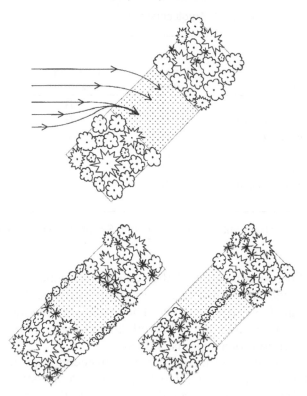

FIGURE 7.2. The top drawing shows side swirl caused when wind is channeled between blocks of tall, relatively impermeable vegetation. The two bottom drawings show layouts used to counter swirl.

Table 7.1 provides a rough guide to what may be expected. Using selected agrotechnologies, this ranking is based on the standard design, with an unprotected primary crop oriented in a way that heightens the effect most, i.e., the wind is from the most damaging direction (e.g., along rows).

## WIND-COUNTERING AGROTECHNOLOGIES

A number of principal-mode agrotechnologies have positive windbreak characteristics and are not susceptible to some or all wind

TABLE 7.1. Agrotechnological susceptibility

| Agrotechnology | Wind stress | | |
|---|---|---|---|
| | Horizontal winds | Lodging or pushover | Wind tunnel |
| Agroforests | R | R | R |
| Alley cropping (hedgerow) | R | R | S |
| Alley cropping (tree row) | R | r | S |
| Living fence (hedge) | R | R | s |
| Living fence (post) | R | s | S |
| Monoculture (perennial) | S | S | S |
| Parkland | S | S | S |

*Note:* R = heavily resistant; r = mildly resistant; S = heavily susceptible; s = less susceptible.

stresses. If properly positioned, they can protect wind-susceptible areas. To provide full protection to vulnerable systems, the designated system should have specific DAPs. The two lower drawings in Figure 7.2 show how pairings are achieved. In this case, two windbreak types are used to protect against side swirl.

The idea behind pairing is that an agrotechnology unprotected against one form of wind affect, is located next to a system that counters this effect. For example, a tall shade system that is prone to horizontal wind tunnels may be paired with a hedge system that prohibits lower level air movement.

Favorable pairings are established by placing systems with a specific weakness (e.g., Table 7.1) next to those with corresponding strengths (e.g., Table 7.2). Determining and matching strengths and weaknesses is an inexact science, but the two tables serve as a rough guide.

## Counterpatterns

The use of counterpatterns is a topic examined more in Chapter 9. This is important in wind management, as they reduce susceptibility between adjoining systems. Through the perpendicular orientation of rows, a wind tunnel in one system is effectively blocked by a neighboring system of the same design. This allows two similar agro-

TABLE 7.2. Wind protection offered by various agrotechnologies

| | Wind stress | | |
| Agrotechnology | Horizontal winds | Lodging or pushover | Wind tunnel |
| --- | --- | --- | --- |
| Agroforests | E | E | E |
| Alley cropping (hedgerow) | E | P | G |
| Alley cropping (tree row) | G | G | E |
| Living fence (hedge) | G | P | G |
| Living fence (post) | P | P | P |
| Monoculture (perennial) | P | P | P |
| Parkland | P | G | P |

*Note:* E = excellent; G = good; P = poor. This describes the mature, standard design form oriented to provide the best protection.

technologies to counter a wind stress through layout rather than design modification or the use of an auxiliary agrotechnology. This does not work in all cases.

## Protection Extension

Protection extension is used where a principal-mode system covers such a large area that a neighboring, often auxiliary, system cannot provide complete wind protection. To counter this, one strategy is to select a principal-mode agrotechnology where, for the effect in question and in concert with a properly formulated windbreak and/or adjoining principal-mode agrotechnology, the protection conferred by one system is extended over the area needing protection.

A number of examples can be cited. One is to use a parkland system to counter top swirl. This is used in league with standard windbreak design that cannot extend this form of protection across a wide area (see Figure 2.1 for an example).

Another technique is to plant species that are not prone to the effect in question or where the protection conferred by neighboring structures can be extended over the full area. An example is to plant a crop that does not lodge and is resistant to drying where horizontal winds cannot be fully controlled. In Photo 7.1, drought- and wind-resistant artichoke is planted in place of more wind-vulnerable crops (e.g., maize or wheat) where the wind tunnel air movement cannot be countered. Here countermeasures are limited by the need to maintain a corridor free of tall vegetation.

Along a similar line, different varieties might be employed as a countermeasure. For example, a lodge-resistant variety is planted in the wind-prone center of a large plot.

## Temporal Considerations

As perennial blocks are harvested, especially those that afford wind protection, others must reach a stage where they can fill the gaps opened in the overall protection. This requires some planning. The wind tunnel example (Photo 7.1) shows well-protected crops along the cut. The series of intercropped poplar and maize and larger blocks inhibit wind penetration and damaging effects.

## AUXILIARY SYSTEMS

Of the auxiliary agrotechnologies, only the windbreak is specifically formulated for wind protection. As such, this is discussed in a separate section. *Cajetes,* catchments, infiltration barriers, firebreaks, and water channels offer no appreciable wind protection and are not discussed further in this context. The others offer varying degrees of protection. In a dual-use mode, these can be modified and used as wind control additions.

## Biomass Banks

As features in landscape, biomass banks can be positioned to provide some wind protection. In their taller form, they can be closely interspersed with crops (e.g., as strips) to confer broad protection. There are pure placement opportunities where, located in wind-prone areas, they still serve their primary purpose, allowing crops to be located in less vulnerable areas. For example, a biomass system for green manure may be placed on a small, unprotected hill where crops would suffer from wind-induced drying.

## Living Fences

As a landscape component, living fences (hedge form) offer some placement opportunities and, where design flexibility exists, they can double as windbreaks. The post type is susceptible to wind tunnel and plots should be designed accordingly.

## Riparian Defenses

Tall riparian defenses can have a windbreak function, either alone in narrow valleys or in conjunction with water breaks in wider expanses. They may be flexible enough to accommodate a third design purpose, as a windbreak. This is especially true for fingered and hand-and-arm designs on hillsides or in flatter streamside locations.

## Water Breaks

Another agrotechnology that has the potential to serve a dual purpose, without compromising either, is the water break. It is flexible enough so that a multispecies windbreak is ideal as waterbreak. Placement favors closer interbreak spacing, which does not run counter to any windbreak function. The only proviso is that the water break be perpendicular to the water flow, and this may not always block prevailing winds.

## WINDBREAKS

There are well-established parameters for windbreak designs and, as a separate dedicated agrotechnology, windbreaks should conform to these design ideals. Other agrotechnologies can serve a windbreak function and may be redesigned with the serviceable parameters of the base agrotechnology with compromise to the original design purpose. Any windbreak should be positioned so that prevailing winds are effectively blocked and changes in wind direction are countered.

## Parameters

The primary function of a windbreak is to eliminate ground-level wind. The horizontal effects of a windbreak are in a direct relation to the height of the structure, where the effect at ground level is roughly 20 to 25 times the windbreak height. The second requirement for a windbreak is not to produce any wind tunnel air movement.

Also important is density. At the upper reaches, there can be considerable force pushing against the structure, and this force should not be acting against fewer plants. In addition to pushover, this can result in tree breakage. The upper level should allow some air movement through the structure, reducing the top swirl from a totally wind-impermeable structure.

## *Dedicated Structures*

Dedicated windbreak agrotechnologies come in two basic forms: single row and multiple row. For the single row, tree species should have all the necessary DPCs and DAPs. The rows are tall narrow canopies that are thinner at the top and denser at the lower levels. Another characteristic is a strong, deep root system to reduce blowdown potential and strong stems and branches to reduce breakage.

A number of species have been employed. Specially developed varieties of poplars are widespread in humid temperate regions, and in drier temperate regions, varieties of drought- and insect-resistant cedars find use (e.g., *Chamaecyparis thyoides, Juniperus communis, J. drupacea,* etc., especially *Cupressus sempervirens*). The advantage of a single-row windbreak is that less land is taken out of production.

A second option is to use multiple species in multiple rows. Each contributes to the desired characteristics. One row and one species provide the dense understory, a second row and species provide a less dense intermediate level, while the upper, emergent species is less dense than the shorter-stature component species. A possible fourth row is emergent, disrupting upper-level wind flows. Other species contribute biodiversity and associated gains. With three or more species, the pushover potential is reduced as the base is wider and the roots are interlocked.

Multiple-species windbreaks come in various forms. As described, three species are often employed, two are possible, and more than three are less common but, area permitting, quite feasible. Because each species has specific attributes, each contributes to and helps define the windbreak role.

## Types and Placement

Although species composition is site and climate dependent, the placement of a windbreak often dictates the type used. Windbreaks on the tops of hills or along mounds or ridges are generally multiple row. This is to counter wind tunnel and toppling, common drawbacks in single-row structures. Multiple-row systems are also prevalent in wide fields where winds move along the ground.

Conversely, single-row structures find greater use in valleys and depressions, where fields are narrower, lower-level winds are less, and where soils are generally deeper and anchor the trees better. Outside these guidelines, individual windbreaks can be part of a larger series of structures where the location of individual windbreaks is set by the more dominant shelterbelts.

## SHELTERBELTS

The windbreak can be one component in a larger set of wind structures that span a large area. These are networks of wind-controlling systems that find uses in flat or moderately hilly regions where constant breezes rob fields and crops of moisture. They can be natural plains or sizable denuded areas that will support, often with some establishment effort, tree growth.

Shelterbelt networks have two components: (1) primary structures made up of actual shelterbelts (also called timber belts), and (2) secondary structures, the windbreaks as described in the previous sections. They work together to provide a pattern of wind control for groups of farms or on a regional basis, and these networks can be made up of many windbreak designs, including hedges, to maximize the overall effect, while being land and labor efficient.

## The Shelter Network

Being the primary element in the network, shelterbelts are located perpendicular to prevailing winds but, as topography and cropping needs are paramount, this is a fairly loose guideline. Figure 7.3 shows a classic landscape arrangement. Shelterbelts can be found as ripar-

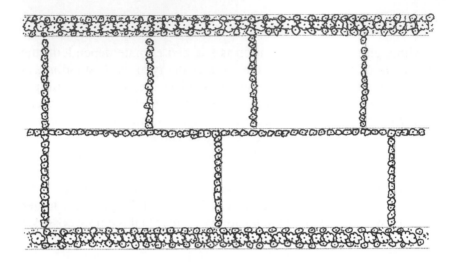

FIGURE 7.3. This classic flatland shelterbelt-windbreak landscape layout shows the shelterbelts (the wide strips at the top and bottom) oriented perpendicular to the prevailing wind. The windbreak components augment or expand the effect.

ian buffers along rivers or streams, used as waterbreaks, and/or be the forestry component within a landscape.

The smaller windbreaks serve to link the larger shelterbelts and/or be contained structures, protecting individual fields. The suggested spacing for the shelterbelts is 50 to 60 heights* apart and, for windbreaks, a 30 height distance (Caborn, 1965).

One advantage of the full system over a few simple windbreaks may lie with increased area efficiency. Rather than needing wider, multispecies windbreaks between fields, the shelter system can permit less intrusive single-line, and possibly more crop-complementary, windbreaks. The bulk of the protection is accomplished through the shelterbelts located at the edge of intensive cropping areas.

As shelterbelts are a major component in vegetative climatic modification, spacing and height will depend on the crops and their suitability for local conditions (Miller et al., 1991). The system can be taller with closer spacing to counter frosts and/or provide a more hu-

---

*Heights refers to the barrier height in relation to vertical distance. For example, if a windbreak is 12 m high, this is one height, two heights would be 24 m.

mid microclimate. Lower heights and more open structure can still block winds but keep humidity closer to that of the local area. Other variations and outcomes are possible.

A brief description of shelterbelt use is provided by Zhang Fend (1996). This example is from a hilly section of Mongolia, where the shelterbelts are in downhill bands connected crosswise by anti-erosion windbreaks. Various species of poplar are employed in these 5 to 10 m wide belts.

Inside these structures, wind velocity was reduced 32 to 38 percent, soil moisture increased 3 to 6 percent, summer temperatures were down by 0.1 to 0.7°C, and winter temperatures increased 0.5 to 1.6°C. The temperature difference may seem small but, if this occurs at a critical point in a growing cycle or serves to moderate temperate extremes, it can be significant. In this case, maize yields were reported 64 percent higher with millet yields up 70 percent.

## *The Primary Element*

The primary elements in the shelterbelt network are the broad swaths of vegetation that cross the landscape. They are usually about 10 m wide, quite tall, and can vary considerably in composition and design (Caborn, 1965). In some cases, the width can be as much as 20 m and as little as 4 m (Moore and Bird, 1997).

The height is important, as the effect on neighboring plots is generally measured in these terms. For example, where the recommended spacing is 30 heights apart, if a shelterbelt is 20 m tall, interbelt spacing would be 600 m.

Composition allows for substantial secondary use. The design can be pitched or vertical, can be row formulated (with a different tree or shrub species in each row), use individual placement, or utilize a more exotic design. Figure 7.4 shows some of the design options. The pitched design is more favorable for certain types of agriculture, as it allows short-statured fruit trees to be located along the edges. The advantage of vertical design is that it offers more opportunities for internal forestry (wood production). Other, more exotic types are formulated around special productive and species needs. These bands of vegetation should not contain breaks. Roads or other gaps should not run straight through a tree belt, but traverse the structure at an angle.

*LANDSCAPE AGROECOLOGY*

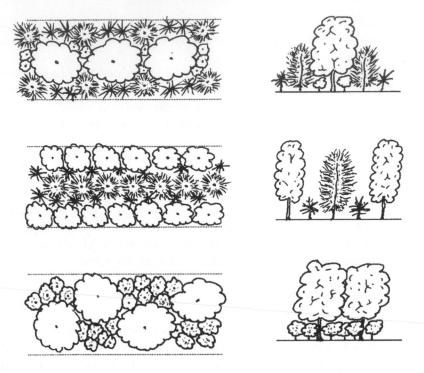

FIGURE 7.4. Overviews and cross sections of some shelterbelt designs. The upper figure shows a pitched design, the middle figure shows a vertical (square) type, and the lower drawing is a more exotic, special-purpose type.

The three suggested design features are (1) biodiversity in perennial species, (2) an uneven upper canopy to disrupt swirl, and (3) slightly wind-permeable structure. The value of the last trait is shown in Table 7.3, where medium dense structure provides the best protection.

## *FROST COUNTERMEASURES*

Where cold air and optimum production is of concern or to avoid total crop loss, countermeasures must be taken (see Figure 7.5). Part of wind movement is the light winds generated by freezing temperatures. In hilly regions, the cold air settles into lower elevations and

TABLE 7.3. Percent wind speed reducation produced by shelterbelt densities at different distances

| | **Percent reduction in average wind speed for the first** | | | |
| | 50 m | 100 m | 150 m | 300 m |
|---|---|---|---|---|
| Structure density | | | | |
| Open | 54 | 46 | 37 | 20 |
| Medium | 60 | 56 | 48 | 28 |
| Dense | 66 | 55 | 44 | 25 |

*Source:* From Caborn, 1965, p. 226.

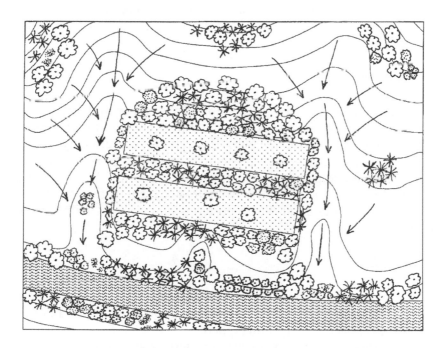

FIGURE 7.5. Two parallel, frost-protected fields (center of drawing). Three forms of frost protection are evident in this overview: (1) the channeling of set-tling frost (arrows show direction) around trees positioned near the fields; (2) the use of high-interface, narrow, rectangular plot shapes with tall borders to reflect heat; and (3) the use of a few parkland species within the plots, also to retain heat.

can affect crops in these prime bottomland situations. Other frost-prone sites are on high plains.

Various countermeasures are used to reduce risk:

1. Canopy reflection
2. Heat sinks
3. Elevation
4. Vegetative decay
5. Scattering

Many of these countermeasures are landscape dependent. Despite the risk reductions provided, cold-tolerant crops in protected locations should provide the last line of defense and an alternate output when all else fails. Other measures include large fans and sprinkler irrigation, which are not part of landscape design and are outside the scope of this text.

## *Canopy Reflection*

Tree canopies over frost-susceptible crops can, by reflecting radiant heat downward, reverse cooling. For this technique, the crop plots are surrounded by dense tree belts. Protection is conferred by having plots of tall trees immediately adjacent to areas to be protected, where the areas are comparatively small in relation to tree height. Wang (1994) reported tea protected from frost by alternating tea-forest strips. Caramori et al. (1996) reported a similar result with shaded coffee.

## *Heat Sinks*

A number of situations have been reported where heat sinks are utilized. They collect daytime heat and, through nighttime release, put this heat back into the atmosphere.

One variant uses stone walls or stone terrace structures, preferably dark in color and facing direct sunlight. Stone terraces in northern Pakistan both support embankments and offer protection from cold mountain air (MacDonald, 1998).

A second variation uses water as the heat sink. It may be associated with aquaforestry or aqua-agriculture or be part of an irrigation system. Traditional agriculture in the Altiplano of Bolivia uses closely spaced irrigation channels to capture and release heat.

### Elevation

Cold air sinks to lower elevations, and farmers take advantage of this. For example, mounds serve this function in the highlands of New Guinea where a 2°C temperature differential can exist between the top and bottom. For a light frost, this can confer the needed protection when the more vulnerable species are planted at the top (Waddell, 1975).

On a larger scale, small hillocks can be similarly planted. If they are circular, they can lead to a ringed planting pattern with the must susceptible uppermost.

A number of other elevation strategies, also used to counter heavy frost, are found in practice. One is to locate frost-prone species on hillsides facing the rising sun. This provides early morning warming while the hillsides shed the freezing air. In cooler climates during the coldest times, the quick warming of tree trunks in direct sunlight can cause splitting. To avoid this, they are painted white to reflect the warmth.

In league with elevation, bottomlands may be protected by providing channels where the freezing air can reach the lowest level without passing over susceptible crops. Air channeling is done with rows of trees along wadis with an open area (a passage) between canopies. Another option is to direct cold air over less- or nonsensitive crops.

### Vegetative Decay

A less well-known cold air countermeasure is vegetative decay. The decomposition of mulch releases heat with positive gain. Waddell (1975) reported that it can raise temperatures 1.2°C. Although rather small, in league with other countermeasures, this can be helpful.

### Scattering

Of greater landscape effect is the scattering of frost-prone species. Light frost can be very uneven within an area and by scattering vulnerable plants some, but not all, of the cold-susceptible species are lost (Waddell, 1975).

This may be one motive for families having dispersed agricultural patches that across a range of elevation zones. Zimmerer (1999) documented this approach in the mountains of Peru and Bolivia.

## DRIFTING SNOW AND SAND

The use of fencing, live or otherwise, can help keep roads clear of either blowing snow (Josiah et al., 1999) or sand (Gaye, 1987). This is a cost-saving measure in road maintenance where fencing is parallel to existing roads (5 to 20 m away) and serves to slow wind and deposit drifting snow or sand, preventing accumulation on roads.

Road protection is a specialized topic, one with landscape ramifications in that fields can be protected in much the same way. Snow deposits, with their hydrology implications, are more appropriate in managing snowmelt for early crops or to place snow drifts where infiltration is maximized. Drifting sand is a different problem that should be kept from agricultural areas with fencing, windbreaks, or some form of sand dune cover.

## FIRE

Fire is a danger and a tool in many landscapes. It can be part of a rotation where burning eliminates crop or fallow residue, or part of active management where burning below fire-resistant trees reduces weeds or kills weed seeds, or where fire is used to control herbivore insects.

Fire can have direct productive purposes. In a documented example from California, native tribes use fire to stimulate and keep basket-making species in a state to provide the raw materials (Anderson, 1999).

Measures to reduce the threat to neighboring ecosystems from the use of this tool are needed. A feature in many forests or forest plantations are firebreaks, and these often define individual plots.

Firebreaks are an agrotechnology, and their use is highly location specific. As a specific technology, the design options are discussed in Chapter 5. Commonly, they are found parallel to roads or railroads where fire danger may be high. Other common locations include ridgetops or ridgelines or along valley bottoms, where an active fire can be more easily controlled.

Firebreaks are usually not placed along the slopes of hills, as a fire can more easily traverse them. However, the use of wide, fire-resistant strips, first as an erosion control measure, might also aid in slowing or controlling low-intensity ground fires.

Usually fire control strips are narrow, 2 to 3 m wide areas cleared of all vegetation. Roads serve a double purpose, as do wide irrigation channels. Vegetation that is fire resistant offers dual use as fencing (e.g., cacti) or as insect control corridors (as discussed in Chapter 8). Vegetative firebreaks expand the number of landscape options.

# Chapter 8

# Integrated Pest Management

A host of organisms can be harmful to the productive purposes of a landscape if their presence is not managed. These can be herbivores (insects, mammals, birds, etc.), animal-attacking creatures (mosquitoes, etc.), fauna detrimental to productive purposes (grain-eating birds, rodents, fruit bats, deer, etc.), or even uncontrolled domestic animals. Plant diseases that curtail production are included. For any one organism, or for a range of natural pests and epidemics, control is best done with a coordinated, planned, and often landscape-wide approach. This is the landscape version of integrated pest management (IPM).

The title of this chapter is somewhat of a misnomer. Generally, IPM applies only to insects and has a monitoring component. Here, it is applied to a range of scourges, while the monitoring component falls outside the immediate scope of landscape design.

The reason for the title is that, in a landscape, IPM has numerous integrated applications that originate from the various countermeasures. Most of the discussion in this chapter is directed toward herbivore insects with some application to insects that injure animals (flies, mosquitoes, etc.)

Efforts are also needed against plant diseases. The ecological countermeasures are less well established and include rotations, barriers, and other forms of spread containment. These are landscape functions.

Various animals can prove detrimental. Birds can eat considerable produce, as can fruit bats, rodents, and large mammals. The control of the small, more mobile pests is a problem with landscape implications. For the larger animals (e.g., elephants love pine plantations and pine bark), hunting or restricted movement is also a landscape solution.

For control measures against any pestilence, be it large or small, some general criteria apply. The NRC (1996a, p. 42) has listed these measures as

1. safety for residents, growers, workers, and consumers;
2. cost effectiveness, ease of implementation, and the ability to integrate the control with normal production procedures; and
3. effective management without negative environmental, economic, or human repercussions.

Within these guidelines, control occurs. It should be noted that pollinating insects, other than honeybees, also inhabit landscapes. This is also a consideration, requiring countermeasures that retain this valuable element while still being effective against unwanted pests.

## BASIC COUNTERMEASURES

Among the range of countermeasures to consider, not all are landscape-wide, but are part of IPM. Most are directed against herbivore insects, while some have implications in controlling other organisms. Where applicable, noninsect examples are provided.

The basic goal is to create conditions agreeable to crops, less so for harmful insects and other pests (Schroth et al., 2000). Interecosystem effects are accomplished through overlapping specific interaction zones, each generating a layer of protection, each focusing on a class of insects, on specific species, and/or on a time period when the insect is most vulnerable or most menacing.

The following list includes some of the countermeasures. They are roughly ranked from the most ecologically friendly to those that have the potential to harm the overall ecology of the land. The ranking is subjective, as this depends on crops, type of insects, climate, and the specific site situation.

1. Habitat control
2. Crop rotations with or without fire to interfere with insect cycles

3. Individual plant resistance
4. Attractant plants (to harbor predator insects)
5. Borders to interfere with insect movement
6. Trap crops (with and without a predator strategy)
7. In situ passive repellent plants
8. In situ repellent plants with traumatic release
9. Microclimatic conditions that favor insect-attacking pathogens
10. Insect traps
11. Cut-and-carry repellent plants
12. Encouraged or introduced large predators (e.g., ducks or chickens)
13. Introduced natural chemicals
14. Introduced artificial chemicals (i.e., commercial insecticides) applied
    • as localized or spot applications
    • over an entire area

With all countermeasures, individual and/or combined, there is little specific information on how these are best used (Hitimana and Mc-Kinley, 1998; Liping, 1991).

## Habitat

Habitat control is one of the most basic countermeasures. It has landscape ramifications in the control of insects as well as plant diseases and destructive fauna. Habitat has a number of facets, both internal and external to a plot. Humidity and microclimate are part of this, both with positive or negative effects on plant diseases (Koech and Whitbread, 2000; Rice and Greenburg, 2000).

Other factors are involved, many under control of the land user, some less than obvious, and all part of integrated IPM and overall land management strategies. An example of a less apparent influence, with broader landscape implications, is the use of natural fertilizer (manure). It has been shown to have a more detrimental effect on insect pests than synthetic fertilizer (Morales et al., 2001).

## Resistance

Among the DPCs are resistance to diseases and insects, including belowground pests. These traits are innate in some varieties and can, through breeding, be brought to others. Along these same lines are the genetically modified crops. They are controversial and have potential ("Much Ado About Nothing," 2002), but are not a complete substitute for the defensive strategies outlined here. Comprehensive defenses work against a range of pests and can slow the process where these organisms gain resistance to or overcome one defense.

## Rotations

Crop rotations provide, at the landscape level, a means to control some insects and pathogens. A number of documented cases have been published. Colbach et al. (1997) have shown that crop rotations reduce the incidence of wheat diseases. Other examples are the control of fusarium wilt on cotton with peppermint (Liping, 1991) and a similar wilt on pigeon pea repressed with sorghum (Rao, 1986).

Fire can be part of a rotation (burning crop residues) or part of a growing cycle (to kill weeds in tree plantations where the trees are fire resistant). As an example, Stringer and Alverson (1994) found that alfalfa weevil eggs were destroyed when, after the cropping season, residues were burned.

## Trap Crops

There are species or plant varieties that herbivore insects find delectable. They can be used to lure insects away from more valuable species. Alone, they provide little long-term benefit as they may only encourage population growth. However, when planted within an insect-predator SIZ, they can increase the effectiveness of both control means. In some cases, they may draw insects to areas where insecticides are spot applied.

These plants may be a useful addition where birds are a problem. A section of early-fruiting plants may serve to keep these pests at bay while the main crop matures and is harvested. In parts of West Africa, villagers plant the rice fields of chiefs or elders before planting their own. Conventional wisdom has it that these early-maturing fields attract the majority of the rice-eating birds, reducing the losses in later planted fields.

### Attractant or Host Plants

Attractant or host plant species encourage predator insects. They provide congenial habitat and/or a nectar source that will lure or retain desired populations. They are not only useful for insect predators, but may be employed, as nesting or roost sites, for insect-eating birds and bats.

### Immunization

Some work supports the idea that plants can be immunized against pathogens and insects. Broad discussion (Day, 2001) and more specialized study (e.g., Ruc, 1990) show that resistance can be increased by stimulating the production of defensive compounds within the vegetative structure of a plant.

### Repellent Plants

A wide range of repellent plants have been identified. They can have a role in countering insects either at the plot or landscape level. Two use options exist. The first is in situ, where these species are grown with productive trees or crops. The second has them conveniently located and cut-and-carried to where they are needed.

### In Situ Repellent Species

These plants are an integral part of plot design. Their passive presence is meant to discourage herbivore insects. If outbreaks occur, traumatic release of contained chemicals can be accomplished by cutting leaves, releasing volatile repellent chemicals, thereby increasing the effectiveness of the plant. The lists of such plants are extensive (e.g., Ellis and Bradley, 1996).

This is not solely an aboveground effect. For example, the incidence of nematodes on plantains was reduced when planted with flemingia (Banful et al., 2000), while leucaena reduced nematodes in maize (D'Hondt-Defrancq, 1993).

This effect also finds application against small animals. Altieri and Trujillo (1987) document the planting of ayacote *(Phaseolus coccineus)* in Mexico, as this plant has a toxic root secretion that discourages rodents.

## Cut-and-Carry

This use of repellent plants does not require the plant-plant complementarity needed when species are grown in close association. Instead, repellent species are grown in conveniently located small blocks or strips near affected trees or crops. When an insect outbreak occurs, biomass is cut and carried to where the outbreak occurs.

Within the landscape, there is no reason why strips of repellent plants cannot double as barriers. In more advanced schemes, in situ repellent plants (strip or intercropped) can drive insects into predator-prey killing zones (see following section).

## Traps

There are some instances where physical traps have proved effective. The examples mostly involve insects that directly affect human health (e.g., flies), but their use in farming and forestry may be beneficial. Traps are aimed at specific insect species and may be more cost and ecologically effective in an outbreak situation than broad measures.

## Predator-Prey

Any number of predator-prey relationships exist. Insects prey on other insect species; a range of fauna preys on insect species; birds and animals prey on troublesome fauna; and humans prey on both insects and fauna. Therefore, the categories are

1. insect-insect,
2. fauna-insect,
3. human-insect, or
4. human-insect or human-fauna.

Humans preying on insects may seem strange, but examples exist (Menzel and D'Alvisio, 1998). Although herbivore insects are plant pests, they can also be a desirable commodity, part of entomo-agriculture, or a supplemental addition to another agrotechnological form.

*Insect-Insect*

Insect-insect strategies can be subdivided into two categories; generalist and specialist (Vandermeer, 1995). These are part of landscape IPM.

*Generalist.* Predator-prey relationships are among the most powerful generalist tools, capable of controlling all types of herbivore insects, and are associated with landscape design. This can be in the form of biodiverse field margins or hedges that harbor a range of predators, e.g., spiders, wasps, ladybugs, ants, etc., where any insect that goes near risks life and limb.

Some insects spread widely, while attractant plants, cover crops, or mulch can encourage wider movement for those that do not. For example, a traditional Chinese technique uses straw within fields to attract and hold beneficial insects (Long, 2001).

There can be some drawbacks to an unplanned or unmonitored predator-prey strategy. The possibility exists that some pests may be protected from predation by specialist insects when the predator-prey balance is not understood and measures taken (Snyder and Ives, 2001).

*Specialist.* Under predator-prey relationships, a number of specialist approaches also exist. Plants can be utilized that specific predators find to their liking. For example, ants will populate plant species that provide a nectar source and, in so doing, increase the morality rate for other insects within their SIZ. For example, weaver ants have been suggested to control small herbivores on citrus in Southeast Asia (Van Mele and van Lenteren, 2002).

*Fauna-Insect*

A wide range of animal species eat insects. Ranging from birds to bats, they can be encouraged, through landscape design, to stay and their task be made easier. At present, there is little guidance on how this can be done and some delicate balances may need to be maintained. For example, there is the problem of encouraging insect-eating birds while discouraging similar fruit- or grain-eating species. A cost-benefit analysis where the losses are weighed against the gains can be applied, requiring subtle changes in habitat, temporal changes in cropping patterns, or entirely abandoning this approach.

Whatever the case, numerous examples exist where this approach has been used successfully. For example, woodpeckers' beneficial effect on borer populations outweighs any bark damage caused (Whitcomb, 1970). A balance may be hard to achieve, but generally, bird populations can have a greater positive than negative effect on crops (Tremblay et al., 2001).

Other cases exist where fauna can be introduced or closely controlled. Chickens eat insects and are used in some cultures to counter herbivore insects. Ducks and chickens can seriously reduce fly populations in barnyards. Insect-eating bats can be encouraged with bat boxes or by preserving nearby cave or other habitat environments. In aqua-agriculture, fish added to paddies have been used to control mosquito larva and disrupt the life cycles of other waterborne insect pests. Frogs can accomplish similar goals against flying insects.

*Fauna-Fauna*

A number of examples of fauna-against-fauna control exist. Cats, predatory birds, and snakes are useful against rodents. These can be encouraged. For example, an early German practice was to place roosting poles (T-shaped poles) in grain fields to encourage birds that hunt mice and rats.

*Human-Fauna*

Human control of bothersome fauna is both a food supplement and control mechanism. These are linked. Hunting as means of sport or subsistence is discussed in Chapter 12. The control aspects are presented here.

Birds and small animals can be pests in that they graze vegetation, trample plants (e.g., deer and geese), uproot crops (to get at in-soil insects), and can eat and foul animal feed (O'Connor and Shrubb, 1986). Outside of hunting, there are other means such as scare tactics (e.g., movement, scarecrows, and noise), habitat control, and the use of trap crops.

For hunting control, a number of guidelines have been proposed (O'Connor and Shrubb, 1986). This may be best accomplished closer to the habitat, rather than where the damage actually occurs. Also, hunting should continue beyond the point where prey becomes scarce. This way, the breeding population will be less able to make up for the losses. Hunting may be less effective than scare tactics or trap crops where population are in transit (e.g., migration) or move great distances in search of food.

## Barriers

Herbivore insects must seek desirable plants, and a vegetative border can serve as a barrier to their free movement. For this to occur, the border must have sufficient height to serve the intended function or have repellent properties.

Beyond the basic parameters, barriers can also serve as a hunting ground for predator insects, a habitat for insect-eating animals, be composed of trap crops to attract insects to predators, or be composed of repellent plants to further hamper insect movement.

## Reservoirs/Corridors

These are landscape features that function as a sanctuary for predator insects and insect-eating birds and mammals. Reservoirs and corridors are used where a predator-prey strategy is employed. Ideally, these areas should favor specific species of predator fauna (insects, birds, or bats) and not be a good habitat for organisms that are detrimental to crops.

Reservoirs can be an area of natural vegetation, an auxiliary structure, or a productive entity within the landscape. A species-biodiverse hedge or grass strip can serve this function, as can a riparian buffer or plantation block. Bio-rich neighboring fields, such as mixed grasses or wide field margins, also serve this purpose. Habitat need not be large. A single, older, well-established tree in a suitable environment has been shown to harbor around six bird species (Herzog and Oetmann, 2001).

The placement and size of reservoirs and corridors also play a role in effectiveness. The specific interaction zone from a hedge or strip, where predator insects are most active, can be about 2 m, although others have found wider colonization (e.g., Alomar et al., 2002). With and colleagues (2002) found reservoirs, transferral paths, and SIZs to be insect dependent, where different species of predator insect function best with certain reservoir placements and corridor dimensions.

This raises the question whether many smaller areas or a few larger ones are best (Tscharntke, 2002). Answering this question implies a good knowledge of the behavior of predator insects and an ability to use it to formulate dimensions and optimal placements of these killing zones. Along these same lines, wind direction and reservoir height may have some impact on the SIZ but, at this point, these design factors are mostly unknown.

Composition is also of concern. Reservoirs can be more effective if paths or corridors are enriched with key attractant species to foster insect spread. Flowering and nectar-producing plants makes these more attractive to predator insects (e.g., Nicholls and Altieri, 2001) and, as a side benefit, to those lesser, often unnoticed, insects associated with crop pollination (Milius, 2002).

Other reservoir schemes may require blocks of natural or less-exploited forests. Owiunji and Plumptre (1998) found heavily logged forests less effective in harboring bird insectivores than unlogged areas.

As part of this, internal agroecosystem plants can expand the corridor-related SIZ. For example, predator-insect favorable cover crops may be part of a reservoir-pathway strategy.

## Insecticides

The use (and overuse) of insect-inhibiting and/or killing chemicals is an established practice in high-intensity agriculture. It is very effective, but with a range of unintended consequences. In addition to instances where misuse has affected human health, it can have ecological repercussions (e.g., water contamination and a reduction in beneficial insects).

## Select Application

One way to mitigate harmful effects is through more sophisticated insecticide application. A number of options exist (Jordan et al., 1996):

1. More specialized, less invasive chemicals
2. More frequent use at minimal dosage
3. Spot applications in and around outbreaks
4. Timing that coincides with local outbreaks and/or emergence
5. Granules that slow release
6. Low-volume sprays

Spot application means that chemicals are directed where they are needed. For example, if a species of insect must walk up a tree stem as part of its life cycle, a band of chemicals, applied around each stem, may inhibit movement and effectively eliminate the threat. The case study at the end of Chapter 10 promotes the selective use of insecticides in a transformation to more ecologically friendly agriculture.

*Broad Use*

Insecticides have use in agriculture, but the application of broad-spectrum chemicals across the landscape is not regarded as an environmental plus. There are a number of factors to consider. Not all insecticide use confers a long-term decrease in insect populations. Van der Valk et al. (1999) found that an increase in grasshopper populations occurred after the immediate effects of a broad-spectrum insecticide subsided. This was because the positive predator-prey relationships were destroyed.

The same can occur when spray drift affects unsprayed neighboring plots. It can accentuate the negative, as enough insecticide may reach these adjoining plots to destroy predator-prey relationships, but not enough to completely eradicate the herbivore insects. Drift can be countered to some degree through more controlled application and thick, between-plot tree barriers. Without controlling measures, e.g., barriers, when one plot is sprayed, neighboring plots must soon follow. This can force all land users into using the same control standard.

Not all insecticides have the same environmental impact and should be treated accordingly. Certain chemicals, such as DDT, have shown considerable negative aftereffects, and there are widespread legal bans on their use. Other insect-countering chemicals are less toxic.

Among the most benign is diatomaceous earth (Pest Control, 2001). This mineral has proven effective in controlling a broad range of insects but, since it is also used as a food additive, does not directly harm fauna. Despite the benign characteristics, it can still ruin a predator-prey strategy.

A range of mild chemicals exist, some acting against pathogens. For example, to counter crop mildew, cow's milk has proved effective without environmental harm (Bettiol, 1999).

## COMBINED STRATEGIES

The control of insect pests, either herbivores or those attacking fauna (e.g., mosquitos, flies, etc.), can be formulated upon different overlapping strategies. Part of this is rooted in the larger landscape, while other elements are added through agroecosystem design.

IPM is well established in agriculture. The earliest efforts were directed toward the thoughtful use of broad-spectrum insecticides. The goal is to apply these insecticides only when insect populations reach critical thresholds based on population ecology and the projected increase in the population.

Insecticides are not the only method to control insects, and later developments in this field have begun to look at insecticides more as a tool of last resort. Among the developments are the generalist-specialist approach, where measures are taken to maintain overall herbivore insect control and, when outbreaks occur, temporary countermeasures are invoked.

The overriding principle of agroecological IPM is to put the most environmentally friendly control methods first. Stronger countermeasures are used first against minor outbreaks and, as a final resort, the more temporary but more severe countermeasures are invoked against major outbreaks.

### Generalist-Specialist Strategies

Using the generalist-specialist approach (Vandermeer, 1995), certain countermeasures are always in place and functioning. As some countermeasures work against a range of insects and others are more specialized, if this strategy is chosen, a mix of measures is better. How many are used depends on the amount of risk that needs to be contained. The idea is to keep herbivore insect populations, through basic countermeasures, below the level where serious crops losses occur.

A specialist strategy is activated when populations of one insect species reach a level where economically significant damage can ensue. This triggers appropriate countermeasures, either designed against the one species, or a stronger, broad-based approach.

## Tracking-Ordering

Figure 8.1 shows the basic relationships between the stronger, immediate, and more ecologically intrusive countermeasures and those that are less potent, less forceful, more passive, and more environmentally friendly. The idea behind tracking or ordering is that there are generalist and specialist tracks.

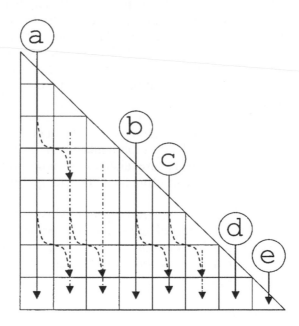

FIGURE 8.1. The basic idea behind tracking-ordering is expressed diagrammatically. The vertical solid lines (labeled a, b, c, d, and e) are generalist tracks; the offshoot (dotted) lines are specialist tracks. The upper squares are the more landscape-oriented pest control measures. The lower squares represent plot or spot applications. The countermeasures on the left side are more benign; those on the right are more immediate and severe (hence fewer are needed).

The different tracks are a series of integrated insect countermeasures (the blocks in Figure 8.1) where each added defense supplements, but does not interfere with, the previous planned and executed defensive measures. The topmost are those fixed in place (and in large part of landscape origin) and presumably less costly. At the bottom are those that can be utilized in a specific location (spot application) and with less notice.

There are some tracks for generalist insects (Figure 8.1, labeled a, b, c, d, and e) others for specialist herbivore pests. A track is selected, either benign or severe (Figure 8.1, left to right) and these are implemented from top to bottom following each suggested column progression. If the generalist insect danger threshold (as determined through population dynamics) is not reached, the top-to-bottom progression may stop at an appropriate point or continue the full counter sequence.

If a specialist insect species attacks, the switch can be made to another track (vertical dotted lines, Figure 8.1). The primary or generalist track can be abandoned when switching to another more specialist track or followed if other threats may be upcoming.

At the right of this grid are the stronger countermeasures with less need for subsidiary or supplementary measures, hence the shorter columns. In the taller columns, each countermeasure may be less effective, so, to accomplish the task, a wider range is needed to augment or fill defensive gaps.

As examples of first-column defenses the blocks (Figure 8.1, left) may be

- landscape, conducive to predator fauna,
- movement barriers,
- predator-prey hedge rows or grass strips,
- cover crops or mulch to promote predator movement,
- repellent plants located so as to drive insects toward hedges,
- cut-and-carry strips for outbreaks, and
- introduced predator insects.

A midcolumn defensive series against a specific pest can be

- spread-prevention barriers,
- bird and/or bat houses to retain unique predator fauna,
- trap crops to lure and concentrate a specific herbivore insect,

- host plants to retain predator insects, and
- introduced predator insects to supplement natural populations.

A second to the last column block series can include

- interspersed trap crops and
- the spot application of insecticides.

A last column (furthest right) defense would be

- the application of a broad-spectrum insecticide.

It is possible to adopt a multicolumn strategy. It is environmentally better to go to left-column solutions rather than to the right although, as a late measure to save trees or crops, a severe movement right and downward might be advised.

Clearly, the research needed to fully utilize this triangle approach is not in place. Multiple, coordinated countermeasures may be used where one measure is made more effective by a second. As examples, a system can also be designed such that bird movement exposes more insects to predators (free-ranging domestic birds may cause insects to fly more), or trap crops and repellent species may concentrate insects in killing zones, making it easier for chickens to find insect prey.

## LANDSCAPE FEATURES

For landscape IPM, ecosystem design and placement of principal-mode systems is the key, but with the option of plot redesign and auxiliary systems. The latter include barriers, predator insect reservoirs, etc.

### Expansion

The use of internal plants and subsystems to host predator insect species is well documented. This is a prime example of ecological expansion, where the range can be favorably extended through redesign. Adding more biodiversity through predator-insect agreeable species could enlarge the neighboring SIZ. Cover crops, hedges, grass strips, and parkland trees are examples of vegetative additions that exploit and extend predator reservoirs and corridors across a productive plot.

## Augmentation

For insect management, the idea that internal agroecosystem properties are strengthened by the design of adjoining systems is less substantiated, but may be equally valid. Insect control has clear aspects and some that are less acknowledged.

Indirect control mechanisms, those that result from fauna (e.g., birds and bats), are a corollary of a favorable habitat, and, if neighboring, a feeding zone can extend a considerable distance from a host ecosystem. For predator insects, the picture is less clear, e.g., although winds may serve to extend the SIZ in question.

Another aspect of ecological augmentation is the island hypothesis (Stocks, 1983; Gliessman, 1998). With this strategy, destructive insects are kept away from a delectable crop through a surrounding belt of disagreeable vegetation.

Although not fully understood, it is found in practice with documented examples. Platt et al. (1999) found a buckwheat strip effective in controlling the spread of cucumber beetle. In forestry, similar arrangements have been shown to work in tree plantations (Bragança et al., 1998). The island hypothesis has also been alluded to in connection with crop-destructive small mammals (Beckerman, 1984).

## THE AGROTECHNOLOGIES

In contrast to other landscape threats, insect and disease control is, in large part, best internalized in principal-mode systems. Part of this is that there are few dedicated auxiliary systems and, except for the vegetative island hypothesis, the insect-regulating effects of adjoining systems only extend for a short span. The role of the agrotechnologies, through internal dynamics, extension, and augmentation potential, may prove helpful in placing this in a larger pest control context.

Auxiliary systems offer the same opportunities as principal-mode systems. The only difference is in the amount of landscape flexibility. Shorn of the need to provide output, they can often serve a double duty. In addition to their primary task, through vegetative modification, they can be converted into pest control systems.

## INSECT COUNTERMEASURES:
## A MEDIEVAL CASE STUDY

In the middle ages of Europe, a landscape evolved that fit the ecology of region and the crops raised (see Figure 8.2). The physical landscape has mostly disappeared; only vestiges remain, but it has resurfaced in parts of Africa in a remedial form as alley cropping.

This landscape is based on long, narrow, irregular fields separated by hedgerows or bordering strips. They have a practical consideration; in plowing, draft animals are more efficient if they turn less often.

The hedgerows, through collection and release of nutrients by way of above and belowground biomass, have positive effects on soil fertility. More important is the planned rotational sequence. The planting sequence (Vasey, 1992, p. 160; Gras, 1940, p. 37) was winter grains (wheat or rye with more rye than nutrient-demanding wheat), spring grains (oats, barley, broad beans, vetch, or buckwheat) and fallow. The fallow is 12 to 14 months with repeated (often three) plowings to infuse captured nutrients into the soil structure.

Crop rotations, mixed-species hedges, fallow lands, field margins, scattered parkland trees, and long narrow fields serve as a mutually reinforcing set of insect control measures. Less obvious are the externally influenced pest control measures rooted in the larger landscape. The village contains structures conducive to insect-eating animals. Barn swallows, owls, and bats dwell in old, loosely constructed churches and barns. The trees, fruit and shade, also offer habitat opportunities for predator fauna.

The village is at the center, surrounded first by the gardens, then a belt of fruit and nut trees. Outside this are crop fields and, outside of this, forests. The forests contribute to the agroecology of the overall system while providing firewood, and serve as a feed source for animals. In these forests, pigs feed on acorns and beechnuts, cattle on grasses, goats on coarse vegetation, and horses on the leaves of shrubs and branches, while the forest provides a reservoir of insect-predator animals and insects.

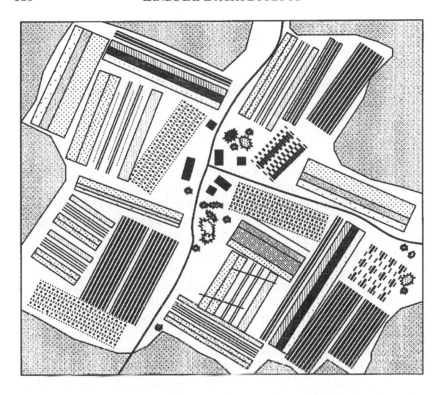

FIGURE 8.2. This representation of a medieval landscape shows long narrow plots where insect and disease control is a function of both the rotational sequence and the interface between the plots and hedges, grass strips, and field margins.

Increased ownership rates of cattle and horses, and the need to graze on fallowed land, made strips unsuitable. The ecological loss in not having a close interspecies interface would have been made up in improved rotations and possibly more resistant varieties.

# Chapter 9

# Patterns (Spatial and Temporal)

In Chapters 4 and 5, the components of the one-plot one-agro-technology landscape are presented. Their placement with regard to some natural events is discussed in Chapters 6, 7, and 8. More can be done locationally to extract the full socioeconomic and ecological potential from these placements. Again, this can address risk factors or start the process of looking at socioeconomic need and cultural agroecology.

As a mix of selected agrotechnologies, the landscape exhibits clear patterns that stem from the view or uses made of the agrotechnologies, e.g., a landscape as a mix of auxiliary and principal-mode systems. These patterns can stem from the use of event-countering layouts (i.e., water, wind, and pestilence) discussed in the previous chapters. The agrotechnologies alone can provide a productive and environmentally sound landscape, but more can be done in arriving at an (agro)ecologically viable landscape. One pattern is important in this regard.

The basic spatial (physical) landscape evolves as a function of a number of immediate influences. Foremost among these are cropping needs, degree of mechanization, and road and other infrastructure placements. Further down on the list, but equally influential, are some less obvious factors including resource apportionment (e.g., labor), knowledge of cropping alternatives, etc. In this context, it is possible to delve deeper into the bag of tricks to increase the ecological gains without altering the above prerequisites.

There is also the temporal landscape and associated patterns. With long-term perennials, a comparatively static situation is presented. This does not preclude temporal features, as there is still the timing of the numerous managerial inputs (planting, pruning, weeding, nutrient inputs, etc). This is an often underused aspect.

With seasonal crops, the transient phases open another entire agroecological dimension. Along with input timing, they offer a wide range of possible ecological benefits.

Within preestablished socioeconomic and agroecological parameters, some clear landscape agrotechnological types evolve. Once the agronomic and ecological motives are explained, what remains is a cultural outgrowth.

## AGROTECHNOLOGICAL LANDSCAPE TYPES

Based on the agrotechnologies described in the previous chapters, some distinct landscape types have emerged. Their use, whether they are (1) ecologically independent, (2) systems acting in unison, or (3) a combination of types, forms the base for various landscape designs. The expanded list of landscape types is as follows:

independent agroecosystem,
mutually reinforcing,
supporting auxiliary,
aggregated landscape,
rotational-taungya, and
mixed natural ecosystem.

This does not exhaust the possibilities but, given the range of alternatives, only the common variants can be presented here.

### Independent

A landscape can be viewed as a series of separate, principal-mode, self-sufficient agroecosystems where productive and environmental problems are handled (1) either through imported resources that do not originate in neighboring plots (e.g., chemical fertilizes) or (2) through internal agroecological dynamics with or without rotations and fallows. For the first case, imported resources commonly support monocultures (as externally funded plots) and other high-yield systems. One goal in agroecology is to promote internal agroecological self-reliance (without environmentally harmful chemical inputs) through planned biodiversity, and many of the agrotechnologies achieve this in some form.

Internal plot sustainability is championed as one of many viable options. The alternative, a landscape that achieves broader objectives, is more the focus of landscape agroecology.

## *Mutually Reinforcing*

A landscape comprised of set principal-mode agrotechnologies can rely upon interplot dynamics to resolve productive and environmental problems. The tools used are the type of agroecosystem, the DAPs, the size, the shape, and the location of each plot. Through these variables alone, a mutually supporting landscape can be secured. An example provided earlier in the text has wind pairings where strong protective attributes in one system are matched with susceptibility in a neighboring system (see Figure 2.1).

## *Supporting Auxiliary*

Auxiliary systems, rather than interplot, principal-mode effects, can achieve agroecological objectives. The principal-mode agroecosystems can possess some of the needed internal agroecological dynamics, but the internal effects can be augmented or, if missing, supplied by auxiliary systems. The tools used are type of auxiliary system, location, and pairing, along with size and shape of recipient plot.

## *Aggregated Landscape*

A composite landscape, containing a mix of autonomous, mutually reinforcing, and auxiliary systems, is also a possibility. This mix may offer the best opportunities for attaining an ecologically holistic form, but only with a high degree of biodiversity and agroecological complexity.

## *Mixed Natural Ecosystem*

This landscape type relies upon strips or blocks of natural vegetation to supply agroecological effects that are missing and needed. These systems serve much the same function as auxiliary systems, but offer more opportunity for the natural ecosystem to function within productive areas. For example, birds, insects, bats, and other fauna can live and thrive within these areas, possibly providing some agroecological benefit to cropping systems.

### *Rotational—Taungya*

Temporal landscapes are a specific type, formulated around the use of, and the ecological dynamics associated with, temporal agro-technologies. These include crop rotations, overlaps, taungyas, and/or age sequences. These require temporal perspective to achieve the needed economic and agroecological dynamics in a constantly evolving panorama. The primary species can be a tree crop or forestry species or contain some mix of species, either within or between different age blocks. The farming activity within the rotation or taungya phases can supply or augment biodiversity, where interplot effects are used to reinforce or provide needed agroecological dynamics.

These contrast with those described previously, but are not as exclusive. They utilize many of the same dynamics found in static versions, but with more emphasis on unique temporal occurrences (e.g., insect control through rotations or wind control through nonperennial vegetative placement).

## *THE PHYSICAL LANDSCAPE*

In the previous section, some broad landscape types are described. To arrive at or formulate a specific type is more involved than simple descriptions may suggest. The complexity is in the available options and details. The actual placement of plot-based systems is highly dependent on topography and should confirm to that suggested, e.g., the placement of water management systems (see Chapter 5). Flat sites offer more options than are available with hillsides, hence more operational complexity.

### *Flatland*

In not requiring an ordering of erosion control agrotechnologies, level or moderately sloping sites have a greater number of options. Plots can be placed in sequence or utilize other more exotic placement patterns. In the normal sequence, as one crosses a landscape, one plot or auxiliary system is located next to another, usually with straight boundaries. This is a block arrangement.

*Block Arrangements*

Level sites divided into square or rectangular blocks are commonplace worldwide. Some of these have unique patterns, many which have not been studied but, in league with other interface dynamics, rotations, etc., could yield ecological and productive advantages. Equally likely, some patterns could offer better road placement and/ or more economical plowing and harvest patterns for farm or forestry machinery or for draft animals (as with the medieval case study in Chapter 8).

Figure 9.1 shows some different block patterns. All have the potential to provide predator insect reservoirs and corridors, windbreaks, or any other ecological structures. The smaller the block, the greater the ecological potential, keeping in mind that small size can impinge upon productive and economic efficiency. These only sample the possibilities.

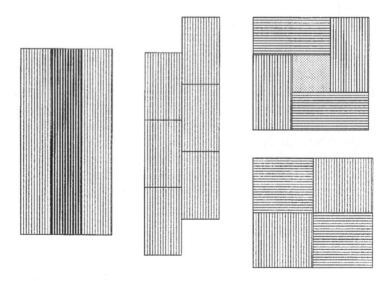

FIGURE 9.1. Various field layouts where, depending on species, relative height, rotation, and topography, each has pluses and minuses. For example, the upper right pattern can confer the advantages of a circular configuration (see text).

*Circular Configurations*

Some landscapes are formulated in a circular arrangement. Pivot irrigation with center well and rotating sprinkler mechanism provides, through default, this form of plot. Often monocultural, these patterns can be internally ringed, where crops follow in progression from the center outward.

Outside of an infrastructural impetus, ringed landscapes may exist for a number of ecological and practical reasons. Stocks (1983) has listed a number of motives for concentric circle plantings, which are enlarged upon here.

1. Planting rings provides maximum exposure to sunlight when the tallest crops are in the outermost rings, the shortest in the center.
2. Having taller flora on the edge of a smaller plot may offer wind protection when the less vulnerable species are in the less exposed core zone. For larger cropping areas, the smaller, more susceptible crops may be better in an intermediate, protected location, shielded by a tall inner core and tall outer ring. The circles and curved rows also serve as a counterpattern to winds blowing from any direction.
3. Circular plantings protect inner crops from pests (see island vegetative hypothesis, Chapter 8, Augmentation), while the outer ring benefits from predator-prey dynamics. Beckerman (1984) notes that this arrangement protects Peruvian crops from small herbivore mammals (e.g., peccaries) because open space in an outer ring is a barrier to these cover-loving animals.
4. Rings improve nutrient flows. The outside crops benefit from surrounding biomass and, where there is a dwelling in the center, the inner crops receive household organic waste.
5. The ringed structure allows dispersal of like crops across a wider segment of site variation than with sequential plot placement (i.e., narrow ring as compared with square or rectangle) and this change brings about a possible reduction in risk factors (insects, climatic, etc.).
6. With a central dwelling, labor-demanding systems are found more in the inner rings, and systems of less labor need are more distant. This improves labor-use efficiency.

An example from Peru (Stocks, 1983) used peanuts in a round inner plot, cassava outside that, maize in the third ring, and a thin loop of plantain in the outer ring. Another study found cassava in the center, a second ring of sugarcane, and an outer circle of banana (Beckerman, 1983). These patterns lack a central household.

The forestry equivalent, the gap structure, allows for wider diversity in species. It has the potential to promote silvicultural prescriptions and ecosystem objectives (Coates and Burton, 1997).

## Circular Variations

The area surrounding villages or households, when viewed from above, can resemble a target, with each ring a different cropping sequence. On a landscape scale, the agriculture of old Germany used household gardens immediately surrounding a village, a *streuobst* (mixed-species orchard) outside the crop fields and grazing, and outermost, the forest zone. This is also discussed in Chapter 12 under house placement and in the case study in Chapter 8. Although the individual fields can be square or rectangular, they still form a roughly circular arrangement.

## Hillsides

With hillsides, the placement situation is quite different. The recommended position for various ecosystems conforms to the land contours (see Photo 9.1). Although there is considerable flexibility, topography places some severe limitations on the ultimate design.

Chapter 6 explains the hydrology and soil considerations of the hillside in relation to agrotechnology placements. In Figure 6.1, five hillside placements for barrier systems are illustrated.

## SPATIAL INTERLUDES

Within the domain of agrotechnology, relative location of the individual agroecosystems can be a basis for ecological strength, with either independent, mutually supporting, or a mix of agrotechnology types. The external ecological dynamics can be generated through placement of individual plants or groups of plants outside of the tar-

PHOTO 9.1. A well-vegetated hillside with different crops planted in contour strips.

get agrotechnology. This can be with principal-mode or auxiliary systems. The ecological forces generated can stem from a contiguous set of systems, with a specific ecological goal in mind, or from interface dynamics alone. It is the latter, and the types of interface, that is the focus of this section.

### *Horizontal Connections*

The landscape can be composed of neighboring and abutting agrotechnologies. It can be either a series of principal-mode systems or principal-mode separated by auxiliary systems. Examples described in Chapters 6, 7, and 8 include live fencing, strips, and windbreaks.

Most times, where one agrotechnology stops, another begins. Other connecting options are possible; the two described here are overlapping and buffered. These are diagrammed in Figure 9.2.

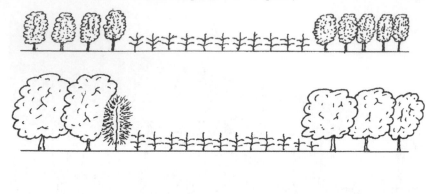

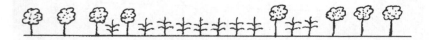

FIGURE 9.2. The three different system interface types are (1) common or simple (upper), (2) buffered (middle), and (3) overlapping (lower). In the middle example, the buffered interface is shown on the left, with the unbuffered and less spatially and ecologically efficient alternative on the right.

## *Transitional-Overlapping*

A transitional or overlapping interface connects dissimilar systems through a spatial overlap. For example, a pasture may connect with a forest plantation by allowing the grass to extend into and grazing to occur among the trees within a plantation boundary (see Photo 9.2). The purpose is to gain the advantage of the edge effect of a taller system to extend both systems. In the pasture plantation example, a SIZ exists where the trees provide shelter from climatic extremes for grazing animals (see Photo 6.1).

An interface can involve an abrupt transition where essentially no modifications in the design of either system occurs. The other alternative involves a reformulation of the agrotechnology to accommodate the needs of the interface. Commonly, this is done through planting density where, with the transition zone, one or both overlapping agrotechnologies may be less densely planted and/or the upper story is pruned. This type of interface implies some degree of plant-plant complementarity.

PHOTO 9.2. A pasture with overlapping characteristics. In this case, the trees are close to the pasture edge, but with wide spacing (as with Figure 9.2, lower right).

## Buffered

Buffer plants are a distinct species used to link different agroecosystems. They are normally used when the adjoining species and systems lack complementarity and a wide, under- or unused space occurs. Buffered designs differ from transitional systems in that the buffer species does not overlap into, nor is it a component of, an adjoining ecosystem.

A buffer species can, by controlling root or branch spread, reduce the intersystem interface distance. Another strategy is to use the interface for facilitative purposes, where a number of gains can occur. Examples include insect containment and wind control (Figure 9.2, middle left).

Another use of a buffer species is to reduce the edge effect in taller perennial systems. The added sunlight can stimulate weed and branch growth along the less shaded edge. Buffer species can reduce associated costs and competition. If the added species has resource complementarity with the two adjoining systems, it effectively connects what would normally be disparate systems. Without or with limited complementarity, buffer species may be classed as an auxiliary system.

## Field Margins

The use of strips (as a course pattern) between different cropping areas has wide application in agroecology. A number of facilitative agrotechnologies (live fencing, windbreaks, water management strips, etc.) champion this layout.

Strips confer, as the primary or secondary purpose, a number of ecological gains. In a partial list, field margins serve

1. as reservoirs and corridors to promote predator-prey insect dynamics,
2. to control wind,
3. to restrain pesticide spray drift,
4. as a refuge for pollinating insects (e.g., Bäckman and Tianen, 2002),
5. to maintain habitat for insect eating and nonagricultural intrusive fauna (e.g., Vickery et al., 2002),
6. for erosion control,
7. as a store of biodiversity (e.g., Ma et al., 2002), and
8. as a sanctuary for earthworms (Lagerlöf et al., 2002) and other microfauna and microflora to repopulate fields.

## Vertical Patterns

Vertical patterns relate to the height relationship between neighboring agrotechnologies or individual species. Height associations can be used as a strategy to promote some associated ecological dynamic, including wind control, microclimate, light apportionment, etc. The two types of relative placement are shown in Figure 9.3. They also apply to the internal structure of biorich ecosystems. This topic is advanced in Chapter 11.

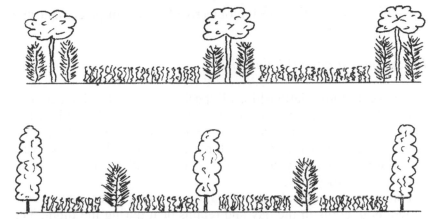

FIGURE 9.3. The cross-sectional relationships between taller and shorter plants. The upper drawing shows minimum interface (or stepped design) where the shorter species is always next to the taller species. The lower drawing shows vegetation arranged in a precipitous or midpoint relationship where shorter plants are always located between two taller species.

## Stepped (Minimum Interface) Relationships

Stepped relationships use a gradient between areas of lower-statured species and taller plants (see Figure 9.3, upper drawing, and Figure 11.2). In simple cases, they function as a buffered interface. In more biodiverse landscapes, patches of taller plants serve other roles, but may still need to be buffered as the height gradient (either buffered, auxiliary, or principal-mode) produces unwanted consequences. One purpose of stepped relationships is to maximize light to light-demanding, short-stature crops where row interface orientation (e.g., north-south) is overridden by topography and associated environmental needs (e.g., erosion).

A medium-height ecosystem, between taller and shorter systems, may also benefit taller agroecosystems, much the same way as gains accrue through the use of buffer species. This may prove advantageous in forestry-crop interfaces, where there are advantages in controlling the negative aspects of an edge effect, including understory weeds and the side branching associated with the additional edge sunlight. This may be a design objective for system placement in a temporal sequence.

## Precipitous (Midpoint) Relationships

In a precipitous relationship, tall species are directly adjacent to areas of short-statured plants (see Figure 9.3, lower illustration, and Figure 11.1). For this relationship to exist, a lot depends on topography and the plot size and shape for the agrotechnologies involved. There are ecological gains or losses associated with an interecosystem height disparity.

Foremost is temperature moderation, where a taller, denser ecosystem both cools in the day and retains heat during the night. The abrupt interface produces a temperature-moderating SIZ. This is due to protection afforded by taller species from the movement of colder air along the ground level and the reflection of heat off taller canopies back to ground level. As a result, one would expect to find more abrupt relationships along river bottoms in arid climates, where frost and radiation of heat can negatively influence crop growth.

Other SIZ effects may be in insect control through predator-prey relationships of birds, bats, or insect-eating insects. An adjoining taller ecosystem may provide favorable habitat and a wider SIZ.

One negative aspect may be light distribution and, where possible, strip orientation can help. As with tree row alley cropping in a north-south strip orientation, midday light is apportioned to the shorter species and, in other periods, more light is allocated to the taller species.

## COUNTERPATTERNS

At times, agroecosystems, because of an inherent property or landscape accommodation, are susceptible to a natural occurrence (e.g., high wind, rainfall). An example is a land contour row orientation used to arrest water erosion where winds parallel the rows and introduce negative crop drying. The solution is a counterpattern where, at set intervals, a row of plants is placed perpendicular to normal rows (see Photo 9.3).

Counterpatterns are also used to facilitate labor movement within agroecosystems or landscapes. For example, for a strip facilitative

PHOTO 9.3. This oil palm plantation has a hedge understory that, in a counter-pattern, runs perpendicular to the row orientation of the primary species (the oil palm).

system, the row within each strip runs perpendicular to the strip orientation. This is used to ease transfer of the biomass to neighboring productive strips. This is shown in Figure 9.4, left drawing, where the vertical strips with horizontal rows (the counter strip) facilitate movement to vertical strips with vertical rows (the productive strips).

On a larger scale, gains might be ecological or involve yields or costs. For example, if the crop on the counter strip is planted first and harvested later, then farm machinery could traverse the longitudinal strip and not need any extra unplanted space to turn around at the end of each strip.

There are a number of design options, as in Figure 9.4, where the center and left drawing contain windbreaks, antierosion or irrigation ditches, or corridors of natural vegetation for insect movement (shown as the darker lines). For each, either internal plot or involving interplot geometry, there are numerous variations in regard to the relative height of the vegetation contained.

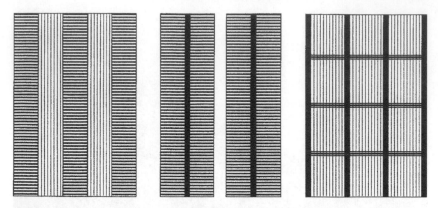

FIGURE 9.4. Some different counterpatterns. The left illustration is designed for labor movement (e.g., cut-and-carry) between the vertical crop strips. The other arrangements are used to block wind and water movement.

## Internal

Within the landscape, there can be a need for barriers within plots that are not an integral part of the contained agrotechnology. Instead, they are an addition, designed for a specific purpose. They could be an agroecological expansion of some positive ecological attribute into a neighboring ecosystem (see Chapter 2, Expansion). They come in different forms.

For the counterpatterns shown in Figure 9.4, the first (left) is a strip system designed to facilitate the movement of biomass between biomass and crop strips. The second (middle) has much the same use with internal hedges to counter insect or wind movements. The final pattern (right) has internal hedges to provide biomass.

## External

Counterpatterns can be established between blocks, either as an auxiliary system or pertaining to the row placement with adjoining blocks. The first option, an auxiliary system, may resemble that shown in Figure 9.4 (left), except that the strips are individual agrotechnologies counterpatterned for the same reasons as internal strips.

The interblock patterns function much the same as windbreaks (Chapter 7) and work best if they share the same design parameters, but, because these are not nonproductive auxiliary structures, they have wider application. They are shown in Figure 9.1, where row placements are such that the adjoining blocks have perpendicular rows.

## *Low Designs*

The low design, as shown in Figure 9.5 (left), utilizes a counter strip or rows of short-statured vegetation. It is used as a soil erosion or hydrology measure designed not to impede foot traffic, the movement of machinery, or light. Within plots, annual or perennial cover crops or low-cut woody perennials (minihedges) can be used. Between low counter rows or strips, there is more flexibility in formulation.

Desirable properties are plant-plant complementarity, nonspreading, nonclimbing, and, if used with annual crops, ease of establishment. If a perennial species is used, it should be able to be suppressed without harm during the cropping period and recover during an off-season or fallow. At times, a cover crop may be the better option or this design may be the first step in establishing a cover crop over a larger area.

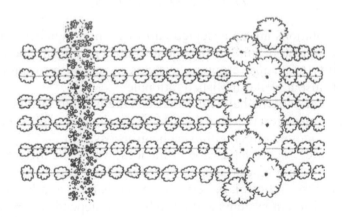

FIGURE 9.5. An overview of two different counterpattern arrangements. On the right, tall trees run perpendicular to the horizontal crop rows. On the left, shorter vegetation runs counter to the primary crop. The latter is illustrated in Photo 9.3.

## High Designs

Where the normal orientation is used as a water control measure or a neighboring ecosystem confers wind protection gains, a high counterpattern decreases susceptibility to wind movement. This type of counterpattern is above the normal height such that foot traffic is not totally impeded. The species used are complementary with the primary crop and a narrow, nonspreading or open canopy.

To avoid channeling wind through the lower opening, this design needs a series of these counterpatterns at close intervals, needs to be used in conjunction with a low-design counterpattern, or needs to be part of a larger shelter system. In agriculture, these are mostly hand-harvested, perennial, principal-mode systems. Where machinery must operate, other measures should be taken.

In forestry, high plantation blocks are commonplace. Here, separately placed vegetative structures in a counterpattern can mitigate any problems caused by the row orientation of the plantation blocks or by large block size. Another option, where the blocks are relatively small, is to vary the row orientation of adjoining blocks to achieve this effect.

## TEMPORAL ASPECTS

Crop rotations are an accepted part of farming practice used, for example, to restore soil fertility and to control pathogens and insects. The ability to change cropping systems in response to market demands is also a well established rationale.

Temporal patterns have microfacets. One is input timing, the sequence of managerial events that can be maximized to great ecological and economic effect within each agroecosystem.

There are broader landscape influences, where the different temporal agrotechnologies (Chapter 5) provide landscape-wide gains in nutrition, insect control, or serve to spread harvest and labor inputs between different systems and periods. A lot depends upon the size and location of agroecosystems, but block arrangements (as in Figure 9.1) are part of realizing or solidifying any landscape-wide rotation gains.

## Input Timing

Within the framework of a rotation, timing can affect the ecology and ultimate acceptability of a system. This is a largely untapped subject, which, upon study, can yield high dividends.

As an example, the scheduling of seed drilling (as with no-till agriculture) can negate some plant diseases. IACPA (1998) noted that barley yellow dwarf virus can be controlled or eradicated in United Kingdom fields by delaying drilling until early October, with the caveat that waiting too long can weaken yields.

This ecological reaction to input timing only samples a long list of such influences. Others include plowing, application of nutrients (either biomass or chemical), insect constraint measures, pruning, uses of fire, etc. If the timing is poor, they can become critical variables. In proper use, they can cast individual agroecosystems and/or landscapes in an ecologically favorable light.

## Rotational Patterns

The advantages of crop rotations for nutrients, insects, and pathogens are well documented, and patterns have been formulated to accommodate these needs. As examples, there may be a need to maintain nutrient potential with the fallow/harvest sequence or there may be a need to reduce nematode populations within the soil through the use of a repellent plant. Less well established are those interplot effects that occur with active rotations in one or more of the plots.

These interplot effects may involve a height differential (e.g., temperate moderation and wind control through height-related rotational patterns); others may use species to evoke specific agroecological principles without a height differential. One hypothesis is the island effect, where insects are prevented from reaching a highly desirable crop by surrounding it with repellent species.

With longer-term perennials (forestry or tree crop plantations), the timing of rotations can be used to modify cost flows or an income stream. The two examples show slightly different stagger patterns:

$$P \longrightarrow |P \longrightarrow |$$
$$P \longrightarrow |P \longrightarrow |$$

$$P \longrightarrow |P \longrightarrow |$$
$$P \longrightarrow |P \longrightarrow |$$

etc....

The sequences can be adjusted, through block size and planting regime, to spread labor needs and to smooth the income stream. With the same total area, the upper situation has smaller planting blocks and uses labor more uniformly, while providing smaller increments of income at shorter intervals.

These plantations can also be semisequential or overlapping sequential patterns where, instead of a clean break between cropping systems, each rotation overruns the next.

$$P_1 \longrightarrow P_1p_2 \to P_2 \longrightarrow P_2p_1 \to P_1 \longrightarrow$$
$$P_1 \longrightarrow P_1p_2 \to P_2 \longrightarrow P_2p_1 \to P_1 \longrightarrow$$

etc....

This formula shows the second rotation starting before the first is completed (i.e., $P_1p_2$). In the case of a forestry plantation, this variation might also be associated with thinning toward the end of a long rotation. The goals are large diameter, high-value logs and to shorten the overall cycle through overlap by using the essential resources freed through thinning. Across the landscape, rotations should be timed such that thinning or other inputs in the different blocks occur in different periods.

### Taungyas

Taungyas exist with forest tree or treecrop plantations, where various understory and/or overstory species are raised in close combination. For the taungya variations (see Chapter 5), intersystem timing can be such as to maintain crop (*c*) outputs within the larger sequence of a tree (*T*) plantation.

This applies to all taungya sequences. A simple version is diagrammatically expressed below. The ending (|) starts the sequence anew with varying-length monocultural phases (denoted with the symbol // ). Within the larger landscape, these sequences are

$$cT \to T \longrightarrow // \longrightarrow |cT \to T \longrightarrow \qquad \text{tract 1}$$
$$cT \to T \longrightarrow //\longrightarrow |cT \to T \longrightarrow \qquad \text{tract 2}$$

etc. . . .

The timing is such that crop outputs occur seasonally. The size and location of different stages is important; size determines when the inputs and outputs occur and placement adjusts the temporal and spatial ecological compatibility of the landscape. Some of the reasons for selecting different temporal agrotechnologies and an overall landscape rotational pattern are listed here:

1. Continued output where, in order to maintain a uniform flow of market or staple crops and more income, a more encompassing strategy is needed.
2. Evenness in labor usage where intense activities, e.g., planting, is distributed across time periods.
3. Temporal—facilitative where each cropping sequence sets the nutrient stage for the next, i.e., leaves the site suitable for ensuing crops.
4. Spatial—facilitative where, in changing crop situations, the facilitative benefits from having one system located near another are preserved.
5. Spatial—complementarity where adjoining systems have plant-plant complementarity between the component species. Keeping these relationships reduces interface distance and promotes more efficient land use.

## Non-Taungya

The overlap in non-taungya systems (pure cropping or forestry) is less common but still encountered. In subsistence farming, this can be part of a productive fallow, where each stage overlaps the next. Other uses are possible.

In agriculture, a drought-resistant crop in the second sequence can overlap with another crop, but be primarily a dry season addition. This can be planted a few weeks before the first (wet season) crop is harvested. This aids establishment. In Africa, cassava is used in this role because of its ability to grow during a dry season.

Non-taungya overlaps may be more common in, but not exclusive to, forestry. With species *a, b,* and *c,* the following sequence may be employed.

$$b \longrightarrow |a \longrightarrow |$$
$$a \longrightarrow ac \longrightarrow c \longrightarrow |$$
$$b \longrightarrow |a \longrightarrow |$$
$$a \longrightarrow ac \longrightarrow c \longrightarrow |$$

etc. ...

Note that species *a* and *c* have essential resource complementarity, while species *a* and *b* may have only partial complementarity. The reasons for this type of landscape can be temporal and/or spatial.

1. This sequence may be a better way to use available soils. Species *a* may draw more nutrients and require the use of species *b* or *c* to replenish the nutrients taken.
2. Species *a* may be in greater demand, and continued supply is needed, but the other species, *b* and *c*, incur less risk or require fewer management inputs. Together these achieve a better balance of costs versus income over the full landscape.
3. Given the complementarity of species *a* and *c*, the overall landscape can be better utilized through the closer spatial association of species *a* and *c* in their mature stages (e.g., smaller plot sizes).

This is one hypothetical example. Any number of permutations can be devised to gain facilitative advantages (e.g., improved soil attributes, fewer on-site insects, etc.) from beneficial pairings along a temporal plane.

### *Vegetative (Fallow) Facilitation*

The previous examples show productive rotation strategies. A distinct temporal agrotechnology also involves fallow usage. As in Chapter 4, there are burn and nonburn fallows, each having landscape uses. The nonfire methods may be better on high-rainfall hillside sites, where the in-place decaying biomass will help prevent erosion. Fire can be used as a tool on erosion-prone areas, where measures are taken to prevent subsequent problems, e.g., cut-and-carry ground-cover biomass and/or strip systems.

As an additional option, fallows can be longer, productive, and more complex. In West Africa a traditional practice is to sequence upland rice followed by cassava and then a mix of productive perennial species. After the cassava, there is a long-term (20-plus years) productive fallow, which is in essence an enriched forest ecosystem with different phases. The latter strategies are land dependent and used only where land surpluses permit very long fallows.

Topography and land area permitting, these fallows have the potential for a more biodiverse, ecologically less-tamed landscape with large amounts of natural vegetation. There are the productive fallows (see fallows, Chapter 5). Over the larger landscape, mixed fallow periods are possible. Usually these are shorter and more intensive on better quality land and longer and less intensive on the more marginal areas. Such a sequence, crop *(c)* and fallow *(f),* may be

$$f \longrightarrow |c \longrightarrow |f \longrightarrow |$$
$$f \longrightarrow |c \longrightarrow |f \longrightarrow |$$
$$f \longrightarrow |c \longrightarrow |f \longrightarrow |$$

where the greater cropping frequency occurs on the better lands. Given the greater flexibility often associated with better sites, gaps in the sequence may be adjusted through agrotechnology selection on these sites. Other variations may be formulated around wet and dry seasons or nutrient needs.

## LANDSCAPE FORMULATION

The previous sections have examined some of the situation patterns that underlie landscape agroecology. Patterns are cumulative in that each level or layer builds to form a final landscape design. These come together in a number of ways, dictated by a host of factors.

### Agroecosystem Size and Shape

In larger landscape entities (e.g., holdings) subdivided by agroecosystems, the size of the subdivisions can be paramount in determining the type and amount of ecological gain. Many small ecosystems present considerable opportunities to utilize interplot and interface effects to achieve productive goals through positive dynamics. This is best accomplished with ecologically compatible systems and large amounts of intersystem interface.

From the opposite perspective, one large ecosystem can also be formulated to achieve the best ecological and productive outcome if the internal design is conducive to and contains sufficient biodiversity to generate the needed dynamics. An extreme example is large cocoa holdings, where the landscape is, in essence, one expansive heavy shade system with considerable overstory and ground-level flora and fauna.

## Cropping Flexibility

The amount of cropping flexibility is yet another means to categorize changing landscapes. This refers to how quickly and cheaply productivity can be switched from one crop to another on a given plot.

From a cropping perspective, seasonal monocultures epitomize flexibility, but flexibility can also be a function of the amount of area containing perennial species, the types of agrotechnologies employed, and their internal content. The type of fallow, either woody or nonwoody, has an effect, as does the number of woody perennials in a given area.

## Cropping Biodiversity

A large segment of agroecological influences in a landscape stems from cropping and noncropping biodiversity. The use of biodiversity can take a mimicry perspective, where the land user strives to duplicate or encompass the dynamics of natural ecosystems over part of or throughout the full landscape. This topic is further detailed in Chapter 10.

Other uses of biodiversity are more modest, only taking advantage of more localized effects starting at the plant-plant interface and the localized dynamics, going to the point where natural dynamics blossom from inherent biodiversity.

A productive entity is a continuum ranging from less biodiverse to highly biodiverse landscapes. The worst case is a large area containing plots of a single clone. Some maize farms in the midwestern United States are of this type. On the other side are subsistence farms where, to provide diversity in the diet and minimize the risk of crop failure, many different productive species and varieties are raised.

## Primary Diversity

In contrast to single-ecosystem agroecology with a mix of primary and secondary species (see Chapter 2), the landscape is often a mix of primary species. They can confer the same benefits without the primary-secondary designation.

This is based on the DAPs of varieties, and even selected niche-diverse clones. As briefly mentioned in Chapter 3, Yoon (2000) has documented an example in China where production of rice doubled when interspersed multiple varieties mitigated the affects of a fungal blast. In another case, MacDonald (1998) reported the use of different wheat varieties in northern Pakistan to counter plant diseases, herbivore pests, and climatic extremes.

## Secondary Diversity

Agroecosystems often contain nonprimary species that have market value, but receive a smaller amount of inputs than primary species. These offer greater opportunity for internal design modification and, as a direct corollary, the possibility for interagroecosystem effects over a larger area.

Because of cost considerations, secondary biodiversity is often formulated around woody perennials. In a changing landscape, while cropping systems rotate, secondary biodiversity remains in place. This has implications in that species chosen and placement used are best with plant-plant complementarity.

## Nonproductive Diversity

Outside of the need for productive species, farms and plantations may maintain areas of natural plants with their associated biodiversity. This may serve a number of ecological functions, but have little or no direct productive role. If these areas are in direct contact with productive units, they can add to or subtract from agroecological gains.

Other nonproductivity comes from the use of individual facilitative plants. The use of facilitative species underpins agroecology, and their uses are broad and many. They can function both within and outside the context of the separate agroecosystems. Interagroecosystem dynamics through individual plants, if used judiciously, can be a positive agroecological force.

Another source of biodiversity is the flora and fauna that are part of occurring ecosystems. Although this component may be lacking in intensively managed monocultures, in many perennial monocultures, such as forestry plantations, the amount of weeds, microvegetation, etc. can be considerable. Although often overlooked, this can exert a strong and positive agroecological force both within and outside an agroecosystem.

## SCATTERING

The discussion in this chapter is mostly predicated on a farm or forestry holding as a continuous unit where the land user controls interplot or interecosystem influences. In a few regions, this is not the norm; instead the holdings are scattered, as small plots or agro-ecosystems, across a wide area. This landscape model has spatial as well as temporal ramifications.

Zimmerer (1999), in studying land use patterns in Peru and Bolivia, noted that discontinuous holdings are an adaptive force that confers many of the same ecological dynamics found with compact and well-designed holdings. The mechanism lies in having sites with greatly varying characteristics. The gains are in

1. managing risk, having crops in various altitudes and micro-climatic zones and not subject to the same localized influences (frost, high wind, soil moisture, etc.);
2. matching the site and crop species, not by finding a species that fits a set soil profile, but in positioning available crop species on appropriate soil types and moisture situations;
3. staggering planting and harvest, where cross-slope temperate or sunlight-inspired differences in germination and growth allow labor to be better apportioned; and
4. allowing the land user to grow a wider range of crops and crop species than possible with a more concise, less variable set of sites.

If there are any disadvantages, it is in increased travel time and in-ability to closely monitor cropping areas for insect or disease out-break or an invasion by crop-eating fauna. Also, if coordination is not maintained with other land users, some possible ecological advantages of intersystem associations are lost. This is especially true where high farming density removes long fallows or sections of undisturbed natural vegetation.

To be fully ecologically effective, scattered plots or agroecosystems should be distributed across a climatically and topographically variable countryside such that each land user has the use of a number of sites with dissimilar attributes. Elevation and aspect (areas facing different directions with varying intensities of sunlight exposure) may be the most common, but other site differences (e.g., soils, soil moisture, drainage) may also qualify and confer similar gains.

As a landscape pattern strategy, scattering is not widely found. If there is insufficient site variability, the existing variability is not well exploited, and/or the areas are widely scattered, labor inefficiencies can negate the ecological benefits.

## THE CULTURAL LANDSCAPE

To this point, most effort has been directed toward variations and design of agrotechnology-based landscapes and the one-plot one-agrotechnology model. This, in itself, is a cultural perspective. Agriculture and forestry often represent the closest association with nature for groups and cultures. Therefore, how the culture views nature is reflected in and manifested through land use patterns. The topic of culture is discussed in Chapter 13.

# Chapter 10

# The Socioeconomic Landscape

The most manifest difference between natural ecosystems and planned and managed agroecosystems is the socioeconomic end goal. This is multipurpose; part economic, part social, while addressing the design package and quality-of-life issues.

All human-managed systems are best understood when there are numerous means to determine (and measure) the efficiency of a system with regard to the efforts and/or inputs expended and what is received. From the agroecological perspective, it is helpful to rank the effectiveness of the DAPs. Evaluation can take any number of forms. In complex systems with more involved agendas, more understanding is gained with each additional assessment type.

Within an economic context, there are various appraisal techniques—revenue, return, and cost—all requiring monetary estimation. Others users, especially those operating outside or with a smaller degree of market forces, may view the landscape in terms of necessities of life, and the landscape can be evaluated in regard to its efficiency in providing them.

Beyond this, how landscape subunits are expressed is a component of the socioeconomic landscape. There are economically independent agroecosystems, where all contribute to the total, without any economically meaningful interactions. These can be agroecosystems that (1) depend upon imported resources (labor, fertilizers, insecticides, etc.) or (2) are ecologically self-contained, relying on strong internal dynamics.

Another option is an agroecologically interdependent landscape where the ecological interactions are, through intent, spread across agroecosystems, and one system depends upon neighboring systems to achieve its purpose. At this stage, the independent agroecosystem approach is much easier to quantify. Interagroecosystem interactions can, when quantified, lead to a higher-order landscape. Which approach is best is still an open question, although, once the principles and practices are in place, the economic gains from a fully functioning agroecological landscape should tilt in favor of more biocomplexity.

In a related topic, the use of bioeconomic models, along with informal optimization procedures, is part of landscape design. Informal optimization is undertaken by all land users, most often without formal mathematical procedures. Although their use may be limited, formal optimization approaches do cast light on alternatives and can sharpen subsequent thought.

Before entering into the economics of a landscape, it is beneficial to review the often overlooked quality-of-life issues. These apply to both commercial and subsistence holdings.

## QUALITY OF LIFE

The elements of subsistence are sufficient water, food, and shelter. Nyong and Kanaroglou (1999) have noted that, where quality water is far from dwellings, locals will use substantially less than recommended and suffer resulting health consequences. Insufficient food and substandard shelter clearly have similar negative consequences. Creating a quality life in a rural setting necessitates access to quality water, a diversity of food types, a comfortable work environment, and other, difficult to economically quantify, but equally worthwhile, landscape attributes.

With market enterprises, quality-of-life gains need not detract from commercial purposes. Through minor additions, these transcend financial criteria, allowing land users to directly utilize the outputs from their land without going through markets. For example, in parts of Spain 50 percent of the herbal medicinal plants are found in local homegardens, and 80 percent are distributed throughout the landscape (Agelet et al., 2000).

Ecological improvement leading to positive quality change can be added to any commercially oriented landscapes, either on the fringes of principal-mode or through auxiliary systems. Multipurpose plants (Chapter 2) are often associated with quality-of-life issues. Other quality-of-life gains stem from the different forms of biodiversity.

Subsistence farmers, as well as their commercial counterparts, live where a more variable, biodiverse, and visually pleasing landscape can offer substantial intangible gains. The rich have country homes to avail themselves of these noneconomic benefits.

A scenic vista is one of the less tangible quality-of-life gains; having a shady place to rest during a hot day is another. A few well-placed trees can accomplish both and much more. Within the following economic discussion, the notion of the noneconomic advantages should not be lost.

## LANDSCAPE EVALUATION

The economic methods to compare individual agroecosystems are presented in Chapter 3. Among them, financial analysis is the most basic. The other methods are the landscape versions of the land equivalent ratio (LER), relative value total (RVT), cost equivalent ratio (CER), and economic orientation ratio (EOR), as well as time value and risk. In modified form, the efficiency ratios indicate how resource capable a landscape is. Time value and risk analysis are less precise, attempting to mirror land user appraisals on key user criteria.

### Financial

Simple financial comparisons have revenue minus costs equaling returns or profits. This applies over the larger landscape where the sum of returns for the existing layout is compared against the potential of any changes.

This type of analysis is fairly basic and is used to point out possible management changes and to suggest economic improvements. The techniques, from economics and accounting, are equally revealing at the plot or landscape level and are well established. Through two select topics, time value and income stream, they are briefly overviewed here. The advantages and disadvantages of a financial approach are presented in the section Cost-Benefit Analysis.

### Time Value

Uneven returns or income from the various landscape components can be evaluated using the NPV calculation (Chapter 3, equation 3.5). On such a broad scale, this is a blunt tool with more insight being provided through an assessment of each individual ecosystem. Still, holdings, especially commercial enterprises, must undergo some form of evaluation, and some inputs (e.g., machinery use) are better amortized across the full landscape.

The same problems that haunt plot NPV extend to a farm or forestry enterprise. Among these is the bias toward short-term gains and less ecologically complex endeavors. Landscape NPV is not without application; it can be useful for looking at the full cost of fallows in land-scarce regions (Ehui, 1992) or as an instrument for studying sustainability within a recognized financial structure (Hansen and Jones, 1996). The latter topic continues under sustainability, this chapter.

*Income Flow*

More telling than NPV may be the income or revenue stream that occurs in formulating and redesigning a landscape. Land users desire an enduring income stream and, if dependent solely upon land for a return, cannot wait an extended period for long-term crops to mature and yield. For example, when replacing seasonal crops with forestry trees over an entire holding, there can be a large income gap after the last crop harvest and prior to the trees reaching market. This gap must be bridged in order for new systems to be accepted or major change undertaken. This works against landscape-wide changes.

*Cost-Benefit Analysis*

Land use decisions have a financial component and involve economic criteria. The ultimate judge on the formulation or reformulation of a landscape may be based on benefits versus costs (either benefits minus costs or as a cost-benefit ratio). This can incorporate all decision inputs.

The decision criteria have four elements:

1. Tangible and quantifiable
2. Tangible and nonquantifiable
3. Nontangible
4. Extraneous variables

The first of these are revenue and costs (R-C), which can, in the NPV form, incorporate time value. The second category is tangible, but more difficult to quantify in R-C form. These criteria use indirect surrogate estimates (or judgments or opinions), which can vary in accuracy. An example is the need for clean water and the value placed on it.

The third category is less tangible and mostly beyond any meaningful financial or economic grouping. Examples are aesthetic (flowers, beautiful sunsets, etc.), comfort (e.g., a cool, shady work environment), or satisfaction (e.g., a pleasantly scented cool evening).

The final group, the extraneous variables (which may be more descriptively called the deadly details) subdivide into social or technical limits. These are essentially yes-and-no criteria on whether the proposed system or change falls within acceptable social or technical limits.

A detail such as an uneven cash flow may doom an otherwise beneficial change. A technical limit may involve an unanticipated soil structure alteration that can halt a highly profitable system. In West Africa, a redesigned system of maize and groundnut failed to produce enough soil biomass to retain moisture, which ended this attempt to improve sustainability (Versteeg et al., 1998).

The details are not always deadly. There can be unanticipated surprises that make a proposed change more acceptable.

## *Efficiency*

The most basic of the plot assessment measures, the LER, can be expanded to multiple plots or agroecosystems. A possible formulation is the landscape land equivalent ratio (LLER):

$$LLER = 1(LER)_1 + 2(LER)_2 + \ldots + n(LER)_n \qquad (10.1)$$

where the individual plots or ecosystems are $(LER)_1$ through $(LER)_n$.

The LER is based on unity and, for a series of plots or agroecosystems, the LLER is similarly gauged. This is done by expressing the parts of a holding as a ratio (1 through $n$), such that these sum to 1.0. For example, where plots make up 50 percent, 25 percent, 15 percent, and 10 percent of a holding, the ratio values are, respectively, 0.50, 0.25, 0.15, and 0.10. Any added and neighboring nonproductive auxiliary structures that take land area but are designed to increase productivity can be considered as (1) a separate system without yield or (2) part of the same system, and any resulting gains are reflected in yield increases for the combined systems.

That is, where plot 3 is divided into a principal-mode (3a) and an added auxiliary system (3b), the resulting add-on change is

$$3(LER)_3 = 3a(LER)_{3a} + 3b(LER)_{3b} \tag{10.2}$$

where area 3 is equal to 3a + 3b and the LER determination for 3a is based on the primary crop yields obtainable for the original un-reduced plot 3. In numerical form for a three-plot monocultural land-scape, the LLER can be

$$LLER = 0.25(1.0)_1 + 0.25(1.0)_2 + 0.5(1.0)_3 = 1.0 \tag{10.3}$$

Where the added nonproductive auxiliary structure (plot 4) is consid-ered a separate system,

$$\begin{aligned} LLER &= 0.25(1.0)_1 + 0.25(1.0)_2 + 0.4(1.50)_{3a} + 0.1(0)_{3b} \\ &= 1.10 \end{aligned} \tag{10.4}$$

For this, 20 percent of plot 3 is taken by the new auxiliary structure, but this results in a 50 percent gain in productivity. The older base (equation 10.2) is used to calculate this change.

Income (LRVT) and cost (LCER) can also be determined in a simi-lar fashion.

$$LRVT = 1(RVT)_1 + 2(RVT)_2 + \dots + n(RVT)_n \tag{10.5}$$

and

$$LCER = 1(CER)_1 + 2(CER)_2 + \dots + n(CER)_n \tag{10.6}$$

There are opportunities to derive economically and ecologically im-proved landscapes with these composite measures. They provide a picture, one best augmented through other indices.

## *Landscape EOR*

In addition to the results from LER and simple financial account-ing, the measure that has the largest impact on landscape understand-ing is the economic orientation ratio (EOR). A landscape is com-posed of a mix of cropping systems, with some intensive input and some less intensive input. These systems often correspond with land type and topography. Understanding these relationships helps to determine economic usage and placement.

Two lines of economic thought are being followed. Some prefer a more uniform overall balance where resources are more evenly distributed between plots. This is based on equalizing marginal gains between plots, and this approach performs best when plots are more or less equal in quality and the crops have similar market value.

In contrast, some land users prefer to put more resources into one or two revenue-oriented plots and less resources into other areas. This fully exploits high-quality land, putting less productive systems and less resources on lower fertility and/or less well-watered sites. The economic rationale for this is that resources are better utilized where the potential return is greatest.

The idea of allocation of intensity is illustrated in Figure 10.1. In the upper figure, more resources are put into a small area. In the lower figure, the resources are spread across a wider area (see also Photo 10.1).

Changes in resource situations accentuate this. When resources are scarce, systems become more cost oriented. This may be a ripple effect, where areas producing core market or staple crops of high value become only slightly more cost oriented. Species of lesser worth become less valuable, produced with few inputs, and the least valuable plants become substantially more cost oriented. This influences landscape design in that systems with the ability to change or modify orientation with relative ease may be best suited for specialized roles and predetermined placements.

Economic orientation affects the transition from a high-intensity to a low-intensity landscape. At lower-intensity levels, the landscape can be resource balanced or unbalanced. At very high levels, the plots are more equal.

## Risk

Many land users depend entirely on productive outputs and have a low tolerance for risk. This is very true of subsistence farmers, who count wholly upon the land for their sustenance.

Economic assessments of the degree of risk are an imprecise science and, as such, may understate the informal, experience-based,

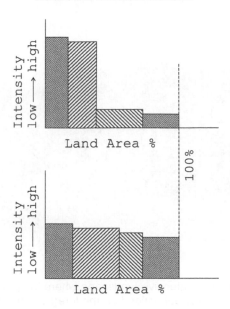

FIGURE 10.1. The intensity balance for two landscape types. The upper graph shows most of the resources going to a few high-intensity, revenue-oriented plots. The lower shows a more cost-oriented landscape where inputs are more evenly applied across the holding.

and opinioned methods used by landowners. Besides land user perceptions, climatic risk factors are often not known or poorly understood, the degree of risk for unmodified or unprotected systems is often not documented, and, in combination, different natural threats may be compounded. Assessment is made more difficult when the risk containment methods for one system spread across the landscape.

There are a number of risk-countering methods, utilizing various agroecosystem and/or landscape modifications (topographical or vegetative). These are often cumulative, where each adds to a total. The costs and a lack of knowledge may preclude use of some options, and the resulting system of defenses against natural calamities may be limited in scope or have severe holes.

A number of examples of risk reduction have been mentioned in previous chapters. Some involve facilitative species (e.g., wind re-

PHOTO 10.1. A medium-intensive landscape where the productive inputs are concentrated on the prime bottomland (in the foreground) while the hillside remains forested and relatively unexploited despite its proximity to dwellings. This photo is from Bahia, Brazil.

duction), soil-incorporated biomass to help retain soil moisture, or imported essential resources (nutrients or water). Irrigated rice may be the most common example, with water channels as a prominent landscape feature.

Of wider use is risk spreading. This is landscape-wide and found in regions where there is one staple crop that is prone to considerable yield variation. The common strategy is to plant a lower-order secondary species that tends to produce some output despite climatic or other natural calamities. This is cheaper to implement, easier to manage, and is found where productive perennial species can be raised. A risk-prone primary species and risk-adverse secondary species can be intercropped or located in separate areas.

Risk spreading also has the advantage of diversifying market risk. If the selling price of one commodity is low, others can be sent to market. The nonharvest option can be employed, where only higher value, marketable fruits or nuts are picked, and lower-value outputs remain on the ground, serving as a nutrient source for subsequent crops. With wood-producing species, harvest can be delayed when other marketable alternatives produce greater return.

## Sustainability

There are a number of views and opinions as to what constitutes sustainability (e.g., Hansen, 1996; Rigby and Cáceres, 2001). These include

1. long-term productivity;
2. the ability of an agroecosystem to defy climatic fluctuations;
3. the use of natural agroecological dynamics or, where that is not possible, the use of natural or environmentally benign inputs; and
4. allowing natural ecosystems to function, including those that promote or sustain indigenous wildlife.

These elements all can be measured using conventional financial analysis, statistical means, or indices with preconceived criteria.

Long-term productivity can be evaluated using NPV, a low discount rate (0 to 2 percent), and a selected present value that indicates when system productivity is falling below set standards. This might include costs of input substitutes or any other fertility alternatives. The advantage is a decision process using conventional analysis. The disadvantage is that this type of methodology can cloud the ecological picture.

The alternative may be to select some future time period and run a simple financial or ratio analysis. For example, the denominator for the LER or RVT is based on a present undepleted or unexploited site, while the numerator uses the future condition.

There is another option where statistical analysis substitutes. A negatively sloping trend line manifests a lack of overall sustainability. The advantage is that a statistical approach removes climatic fluctuations.

The use of indices is far more judgmental than the use of NPV or statistics. The advantage is being able to assess natural compatibility (e.g., wildlife gains), the effect on neighboring systems, and other, less direct productivity concerns. There are dangers in indices. Unless firmly grounded in reality, the values produced may have no bearing on productivity or profitability.

## ECONOMIC CLASSIFICATION

In this section, three agroecosystem types are examined as subsets within a larger holding:

1. Those receiving inputs (labor, fertilizers, insecticides, and/or herbicides) from outside the landscape, referred to here as externally funded agroecosystems
2. Those that are internally self-contained, requiring, in pure form, little more in the way of external inputs than those associated with planting and harvesting
3. Those that are mutually sustained, depending heavily on the surrounding ecosystems, again receiving minimal external resources

There are clearly many opportunities for a landscape composed of a mix of types. Equally possible, and more in tune with landscape agroecology, are agroecosystems that receive some external inputs, generate substantial internal dynamics, and are partially supported through the ecology of neighboring plots.

### Externally Supported Agroecosystems

The concept of externally supported agroecosystems is that the needed resources come from outside the system. A landscape composed exclusively of ecologically nonrelated ecosystems is generally outside the parameters of landscape agroecology. In mixed tracts, those having interdependent and independent ecosystems, the independent plots are economically significant within the larger holding. The distribution of resources to the various systems is an economic question made harder with various ecosystem types and EOR strategies.

### Internally Supported Agroecosystems

In contrast to external-resource-dependent systems, internally supported agroecosystems are generally biomass rich and, through directed design, replace external inputs with internal dynamics. For example, rotations and plant-plant nutrient cycling can replace imported fertilizers and repellent plants, or predator-prey dynamics can supplant insecticides.

Examples are forest tree plots where planting, harvest, and some relatively minor management inputs are the only attention the plot receives. Agroforests are often totally self-contained in a wide range of internal ecological dynamics and, if properly formulated and managed, can be very input free. In economic terms, fully internally supported agroecosystems are very cost oriented.

## Mutually Sustained Agroecosystems

In an ordering of landscape complexity, the highest level and most difficult to analyze are the interplot effects, where the ecological flows are between areas or where substantial cross-ecosystem benefits occur. These landscapes can include the use of auxiliary systems and principal-mode systems incorporating ecological augmentation or expansion.

Also included are management activities that span systems. Active interplot interventions, such as cut-and-carry systems, move plant biomass between areas. The carry labor, as well as the biomass, are external resources.

Mutually sustained systems need not be wholly spatial. They can have temporal aspects based in large part upon rotational dynamics. An example (with guidelines) is given in the case study presented at the end of this chapter.

Greater biodiversity increases economic complexity. This is where the switch is made from systems based on high per plant productivity (Figure 10.2, left drawing) to lower per plant output, but greater planting density (Figure 10.2, right side). One means to handle the complexity economically is through the bioeconomic model.

## BIOECONOMIC MODELING

For the evaluation of internal and mutually sustained agroecosystems, the bioeconomic model can play a role. Computer models are a good vehicle with which to collect and analyze data. During development, models force consideration of the range of ecological forces and, with their more efficient use of available data, have an advantage over pure empirical studies.

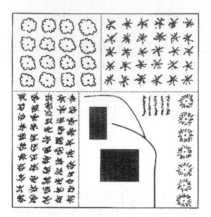 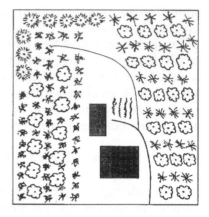

FIGURE 10.2. Two farm holdings. The one on the left has independent ecosystems seeking high per-plant productivity. On the right, the holding is more biodiverse, based on lower per-plant productivity with a greater number of plants per area and heightened interactions between plants and ecosystems. The latter (right) is more difficult to analyze in economic detail.

As data is needed to produce and validate a model, it can guide research along lines not readily apparent. Subsequent studies can be used to further refine or expand an existing model rather than just adding another layer to the literature. After development, a model can be used to examine unstudied variations and eliminate, before entering expensive field trials, variations deemed not economically viable.

If any disadvantage exists, it is where models, due to a lack of user training, are not universally approachable, cannot be readily modified for other situations, and/or are too specific for wider use. There are some good examples of widely adopted single-crop models. For example, the CERES models cover a range of crop types and have a large supporting database (e.g., Jones and Kiniry, 1986; Ritchie et al., 1991).

To be effective, a model must accurately represent what is happening within an ecosystem and produce results that are in line with field data. The workings need not exactly reproduce the full complexity of an ecosystem. A model can be effective if the major influences are captured mathematically. For example, if competitive partitioning accounts for 75 percent of biculture ecosystem interaction with another 15 percent occurring through facilitative effects, these limited interactions will most likely suffice when used to predict the effects of ecosystem change.

There is no reason that models cannot be formulated using DAPs or some other accessible ecological criteria. To be totally effective, any model should have an optimization feature, i.e., a mathematical progression or iteration procedure that leads to an optimal landscape design.

## LANDSCAPE OPTIMIZATION

Economic optimization for landscapes composed of independent agroecosystems is not a well-developed concept and, for interdependent landscapes, nonexistent. Wojtkowski (2002, p. 317) has discussed six categories of optimization:

1. Fixed plots with fixed agrotechnologies
2. Variable plots with fixed agrotechnologies
3. Fixed plots with varied (formulation) agrotechnologies
4. Variable plots with varied agrotechnologies
5. Fixed crop plots with varied selection and placement of shrubs and trees
6. Nonplot with varied selection and placement of crops, shrubs, and trees

These categories are listed in order of analytical and optimization difficulty. The use of fixed plots containing a single agroecosystem (independent systems either externally funded or self-contained) is a simplifying factor. Further complicating any analysis are the variable plots and, at the extreme, nonplot situations. Published examples exist for only the first few cases.

### The Process

With any single-objective landscape, simplicity can be better. For example, where maximizing revenue is the only objective, a single-plot single-agrotechnology landscape containing only the highest-value crop would most often trump other designs. Of course, there are exceptions, e.g., the use of a facilitative species, and seldom are objectives as compact where only revenue maximization is considered.

Land users want productivity, economic return, risk reduction, sustainability, and/or environmental compatibility, and expanding the range of options opens more avenues. A complex multiple-objective landscape with a high degree of biodiversity, different agroecosystems, and different interecosystem interactions can be viewed as a multidimensional mountain range. The peaks (optimization points) can be locally optimal (the shorter peaks) or globally optimal (the highest peak).

The more complex the landscape, the more likely multiple solutions exist. There can be a range of globally optimal solutions and a host of suboptimal solutions whereby land users with the same landscape, crops, and needs can arrive at equally viable, even if substantially dissimilar, landscape designs.

## Goals

Optimization is the goal of any land user and, although they share the common goal of fully functional landscapes that meet shared objectives, the blend of objectives differs. Clearly, immediate productivity heads most lists, with other objectives following. The secondary objectives include risk management (insects, climate, etc.), long-term sustainability, and a host of ecological goals.

There are a number of procedures to handle multiple objective problems:

1. Tie breaking
2. Superior solution compromise
3. Limited compromise

The third option seems best where optimization is a judgmental procedure. Where more formal mathematical processes are used, the first two methods can find use.

The first of the multiple objective methods involves tie breaking. This treats each objective as separate case where ties (different solutions that give equal values for the primary objective) are decided by which best addresses the second objective. Any remaining deadlocks are further differentiated by a third objective. This is best in a highly complex landscape with a number of possible solutions.

The other approaches are based on compromise. A land user is willing to meet slightly less of the primary goal if compensated by more of the second- and third-ranked goals. For example, a forester may be satisfied with less annual growth if fire danger is lessened.

In a complex landscape with highly diverse objectives, the trade-offs are not often linear, i.e., a 50 percent gain in one goal results in a 50 percent loss in another. Among the myriad possibilities are, for example, a 10 percent loss that is traded off against a 15 percent gain elsewhere. When all the options are weighed, superior overall solutions can emerge.

In the absence of clearly superior solutions, limited trade-offs may be needed. For example, land users may accept a high degree of sustainability (productivity over time) only if the cost does not exceed 20 percent of immediate productivity. For this, a land user must accept a situation with limited options. This is often the result of a lack of understanding or the availability of species and land-use alternatives. Given the dearth of published examples and the complexity of the undertaking, the use of landscape optimization remains an open question.

## EXPECTATIONS

Of more practical need is to economically categorize the gains from adopting an agroecological approach. The need to employ, and understand, agrobiocomplexity is a barrier to use, especially where the gains from any one intervention may be small and uncertain.

Outside of wholesale modification (e.g., a total holding redesign and a complete shift in agroecological emphasis), change is normally accomplished through a series of small interventions. Seldom does one intervention alter the ecological balance enough to counter all natural stresses. In making alternations, it is hoped that the sum will become greater than the total of the parts. Tracking this progression, and finding which changes add to the sum, may turn out to be the key economic question.

## ECONOMIC TRANSFORMATION: A CASE STUDY

This case study is from England, where the goal is to apply agroecological techniques to modern farming practice (Jordan, Hutcheon, and Donaldson, 1997; Jordan, Hutcheon, Donaldson, and Farmer, 1997; IACPA, 1998). Although not based upon formal optimization analysis, it still provides insight into the economics and expected change in switching to a more agroecologically formulated landscape.

The system switched from intensive high-input agriculture to less intensive natural means. It was not an entirely novel concept as it updated some earlier medieval rotational techniques (see the case study in Chapter 8). What is of interest is that based on net return, there was not all that much economic difference between a more environmentally agreeable system and an intensive commercial activity.

The key agroecological mechanisms are (Jordan et al., 1996)

1. biodiversity increases through cropping sequences;
2. tillage systems that mitigate, through natural means, pests and diseases, help control weeds, enhance soil structure, and capture and retain nitrogen;
3. IPM models that establish thresholds for herbivore insects, diseases, and weeds with appropriate nonchemical or selected chemicals to control the situation; and
4. field margins as an ecological supplement.

This system uses rotations as a natural means to maintain fertility as well as disease control across a range of crops. The ordered rotation sequence is wheat, barley, beans, wheat, fallow, and oilseed rape, starting again with wheat. Other crops may be integrated into the sequence as required. With this progression, fertilizers supplement natural fertility gains.

The field design uses strips of natural grassland uniformly placed between crop strips. Individual cropping areas are 5 ha versus 30 ha before the change was made.

The basis of this change, and the primary source of the predator-prey activity, seems to be related to temporal aspects, with the natural strips serving as restocking reservoirs and barriers to spread, rather than as a direct source of predator insects. A greater use of IPM is needed, with outbreak forecasting as a monitoring mechanism leading to less, and more environmentally practical, chemical use.

A number of guidelines have been developed for how each crop fits within the sequence (Jordan et al., 1996). As an example, those for potatoes are as follows:

1. Potatoes are grown on a site only one year in four.
2. Nematocides are not permitted, but varieties are selected that resist nematodes.
3. Nitrogen supply at planting should not exceed 60 percent of that needed.
4. Persistent, broad-spectrum, and leachable chemicals are not sanctioned.
5. Fungicide use is based on forecasting.

In contrast, the guidelines for maize are as follows:

1. A two-year sequence on the same site is not allowed.
2. A cover crop must be established during the previous winter.
3. Fertilizer timing depends on peak uptake.
4. Liquid manure is applied only through in-soil injection.
5. Persistent, broad-spectrum, and leachable chemicals are not sanctioned.
6. No herbicide use before growth starts.
7. Intercropping with maize is required in higher rainfall areas.

For other crops, the guidelines are equally divergent. They are the key to this system.

Using the aforementioned guidelines, the overall reduction in chemical inputs was fairly large, with a reduction of 36 percent in applied nitrogen, 26 percent in herbicides, 79 percent in fungicides, and 78 percent in pesticides. As expected in changing from a revenue-oriented to a cost-oriented system, productivity fell by 10 to 15 percent. This was offset by costs that were 33 to 35 percent lower. There was a slight reduction in net return of less than 4 percent.

Mäder et al. (2002), undertaking rotation studies in central Europe, found similar income and cost reductions. Yields were 20 percent lower, while fertilizer use was reduced by 34 to 53 percent. The overall farm income was comparable to more conventional, less agroecologically oriented farms.

Given the overproduction in the agricultural sector in Europe and the abundance of underused land, this trade-off may be well advised.

Although the full range of possibilities was not pursued (including more emphasis on spatial layout and the agroforestry and farm forestry options), these cases certainly offer a direction for converting high-intensity commercial agriculture to an equally productive but more ecologically benign form.

# Chapter 11

# Biodiversity

Biodiversity is a facet of agroecology that has broad ramifications in a productive landscape through

1. an agrobiodiversity approach to agrotechnology usage or
2. a landscape design approach with separate principles and practices.

Either can induce an ecologically sound landscape.

In natural ecosystems, biodiversity permits a full range of natural (ecological) processes to take place. They can prevent or mitigate most forms of environmental degradation, cycle essential nutrients, manage water resources (in normal times and through floods or droughts), control destructive organisms (herbivore insects, destructive animals, and plant diseases), and contribute to maintaining natural flora and fauna.

In agroecology, complex, biodiverse agroecosystems can contribute the same processes. Through the natural dynamics that occur in species-rich systems, the total can exceed the sum of the individual plant-plant dynamics. If managed in accordance with agroecological principles, species-diverse ecosystems can be a positive landscape influence that contributes directly or indirectly to farm and forestry objectives.

There are a number of ways to employ landscape biodiversity, either as part of or outside the parameters of the individual agrotechnologies. The difference between agrobiodiversity and naturally occurring plants forces a reordering of vegetative priorities. This is because most native plants are not valuable in productive or facilitative terms and because the productive landscape may not be conducive to the propagation and survival of many wild plants (e.g., Simberoff, 1999).

As stated by Jain (2000, p. 459), "richness in plant diversity . . . is not evaluated merely by the number of species occurring there, but by the intensity of association and dependence of the indigenous communities on that plant wealth." This chapter ends by examining the amplitudes of this association in agroecology.

## *ECONOMIC ADVANTAGES*

The dynamics of natural ecosystems contain effects that can be expensive to duplicate as inputs or in a nonecological context. Nutrient accumulation and insect and plant disease control are examples where it is substantially cheaper to let nature do the work. For some productive ecosystems, it is not always possible to fully employ ecological forces, as nature imposes some rather strict parameters, and operating within them is not always economically viable.

There are far more opportunities to employ complex ecosystems and natural dynamics than found in practice. Land use decisions are often made according to profitability and/or risk, based on the simplicity of analysis without regard to the advantages of biodiversity. For example, biodiverse shade systems are often associated with shade-tolerant crops, e.g., coffee, cacao, vanilla, and black pepper (Nair, 1993, p. 250).

These systems are less productive than their less biodiverse counterparts, but confer economic advantage. The reduction in management inputs (labor and fertilizers) shifts resources, permitting cultivation over a wider area within the farm landscape, or increasing intensity on selected plots. These systems also reduce market risk by lowering the cost of production and can reduce rainfall risk through facilitative pairings and a more water-efficient ecosystem.

Shade systems are not all bicultures, as increased biodiversity accomplishes much the same. Escalante (1995), in a study of Venezuelan coffee systems, found eight biodiverse variations on coffee shade systems: coffee with

1. shade trees;
2. shade trees and banana;
3. shade trees, banana, and fruit trees;
4. shade trees, banana, fruit, and timber trees;
5. shade and timber trees;

6. shade and fruit trees;
7. banana and timber trees; and
8. timber trees.

These principal-mode systems stress a wide variety of ecological dynamics to achieve productive goals.

Within the context of biodensity and biodiversity, the type and amount of labor inputs change (Torquebiau, 1992; Raynor, 1992). These systems are less time sensitive; the timing of labor inputs can be extremely flexible, i.e., a task can be delayed when other activities have priority.

In addition, labor use can be more casual in that activities may be performed, not as designated tasks, but as offshoots of other chores. For example, while picking fruit, the land user may prune one or two trees. Even in passing, some labor input may be implemented. This may overstate the efficiency of such systems but, by appearing to be less labor intensive, these systems are more acceptable.

## CASUAL BIODIVERSITY

Outside of the agroecological considerations, a biodiversity of useful plants can contribute to the quality of life for rural families. In developed communities, the tendency is to purchase what is needed. With subsistence farmers, more of the daily needs come directly from the land.

The opportunities for casual, useful biodiversity abound when advantage is taken of local flora, which can include forest fragments, riparian buffers, hedgerows, or patches of natural vegetation. Most of the useful species found in these sites are not the most desirable productive species, but still add value to the landscape. They include seldom-used herbs; medicinal plants; species with nice flowers; fruits that may not be all that tasty, but can garnish a table; woody species that offer an array of wood products (e.g., decay-resistant poles); and other interesting additions.

## ENHANCED (DIRECTED) BIODIVERSITY

The notion of enhanced biodiversity combines two concepts:desirable plant characteristics (DPCs) and agrodiversity. Environmental gains accrue from having a balanced and active natural ecosystem. In planned and managed landscapes, the same can occur, but to achieve full measure, these gains are best directed toward specific goals. This is the idea of enhanced agrobiodiversity; a combination of plants with express DPCs, are used to build a user-specific landscape.

In other words, this involves building or enriching ecosystems with specific plant species. Among the options is to increase the agrobiodiversity of principal-mode or auxiliary systems, forest fragments, or other parcels of vegetation.

Examples do exist where landscape biodiversity obtained through enrichment is both a quality-of-life issue and source of profit. Cooper et al. (1996) document cases where a large percentage of family dietary inputs come through the enhanced biodiversity of agroforests. Others have looked at medicinal plants in the landscape (e.g., Agelet et al., 2000).

With some principal-mode systems, agrodiversity can be incorporated without problems (e.g., as canopy species in light and heavy shade systems). For others, the immediate effect on productivity (e.g., with staple crops) seems less than ideal. Still, possibilities abound, not always within the productive (ecosystem) core, but more often within the temporal and/or spatial fringe.

### Individual Plot

A number of methods have been suggested to enhance agrodiversity in individual agroecosystems. On individual plots, the effect may be small, but if applied over many plots, it could be significant. The methods suggested (Wojtkowski, 2002, p. 149) are based on

1. hybrid agrotechnologies,
2. duality,
3. enhanced bicultures,
4. merger,
5. transitional designs, and
6. reinforcing agrotechnologies.

The first of these, hybrid agrotechnology, combines features of two agrotechnologies. If this method includes additional species, then agrodiversity is enhanced. The second, duality, involves additional species to increase system flexibility. For example, two different fruit trees allow the land user to harvest one or both and to adjust output of each to meet market demands.

Another method is the use of enhanced bicultures, where more than one primary species is included and/or where the number of secondary species is increased. This may include the use of buffer species.

The use of transitional design overlaps agrotechnologies, forming highly biodiverse zones between systems. As the amount of overlap increases, merger results so that, depending on complementarity, the two systems coexist in the same time and space.

## *Productive Fallow Biodiversity*

With long-term productive fallows, considerable biodiversity can be maintained across the larger landscape. This occurs when, after a cropping period, the land reverts to a fallow. There need not be a clear line between a staple crop and a productive fallow. Once the fallow period starts, a series of crops follow, each adding to the soil nutrient base, until the sequence starts again with the staple crop.

What differentiates productive fallow from a pure rotational system is that each additional sequence has a productive function, is longer in duration, maintained less, intercropped more, with a larger percentage of uninvited species. If this strategy is followed long enough, what results is an enriched forest ecosystem not that far removed from an agroforest.

Within the entire entity, the different stages, with their biodiversity of different species, give a wide range of ecosystem types and plant communities. As are species enriched through low-level management, the biodiversity present in one area can spread to active neighboring plots, ready for the start of the productive fallow stage (Unruh, 1990).

## *Complex Taungya Biodiversity*

Beyond the less complex taungya forms (initial and final stage), these agrotechnologies offer the opportunity for biorich systems when the more complex forms are used directly or bioenhanced. A landscape sequence may develop from the following temporal progression.

$$cT_1 \longrightarrow T_1T_2 \longrightarrow eT_2 \longrightarrow T_2 \longrightarrow | fT_1 \longrightarrow$$
$$gT_1 \longrightarrow T_1T_2 \longrightarrow eT_2 \longrightarrow T_2 \longrightarrow | hT_1 \longrightarrow$$
$$cT_1 \longrightarrow T_1T_2 \longrightarrow eT_2 \longrightarrow T_2 \longrightarrow | cT_1 \longrightarrow$$

For this, the some tree or tree crop species are used ($T_1$ and $T_2$) but the understory cropping varies and includes species $c$, $e$, $f$, $g$, and $h$. As with most increases in biodiversity, this encourages accompanying microflora and microfauna.

## Interfaces

The interface of two principal-mode systems has traditionally been a source of a large percentage of biodiversity, through overlaps, buffer zones, or just by taking advantage of light and other ecological niches present at the fringe of principal-mode systems. Forest fragments show greater biodiversity in these zones (e.g., Malcolm, 1994). In biorich agroecosystems, Jensen (1993) found that 50 percent of individual tree species in Java homegardens were located at the garden edge. This location has the greater exploitable ecological potential that is outside the parameters of set agrotechnologies. Set systems do not take into account the edge effect.

## MIMICRY

The concept of mimicry is that an agroecosystem can be purposely similar to what nature puts on a site, biodiversity included. As a landscape strategy, it has been discussed and ideas proposed, some advocating that mainstream agriculture move in this direction (e.g., Lefroy et al., 1999; Jackson, 2002). Some commercial (domesticated) species may have lost the ability to compete and thrive in a more competitive, biodiverse, and natural environment, but this may be overstated (Michon and de Foresta, 1997).

As examples of mimicry, parkland systems resemble savannahs with soil nutrient gains (e.g., Velasco et al., 1999; Joffee et al., 1999), agroforests look and behave ecologically like natural forests, oaks are a natural succession in pine forestry that can be exploited in overlapping sequential forestry plantations, and aqua-agriculture and aqua-forestry systems can possess the positive landscape characteristics of their natural counterparts.

Using mimicry, there is more than visual likeness. In full application, the individual plots and full landscape take on the physical and temporal characteristics, as nature intended. This is not a far-fetched concept; agricultural and forestry species share niche properties with naturally occurring species, biodiversity can be ensured through plot design, and ecosystem content can be matched with localized sites (topography, soils, and climate).

Even in nature, variation exists. For example, the ancient forests of Europe and North America were influenced by fire, with open ground and grazing animals. Natural African landscapes follow a similar pattern. An agroforested landscape, one mixing trees with crops and grazing plots, takes on many of the same attributes. Fire is used as a tool and/or plowing substituted for this effect.

Although the species may differ, there are a number of advantages in considering mimicry. This approach does not alter area climate, it offers a coinciding habitat to local and migrating fauna, predator-prey dynamics are maintained, and local species are more at home in more familiar surroundings.

## FOREST GARDENS/HOMEGARDENS

The topic of highly bioenriched forest gardens and homegardens has been briefly presented as unique agrotechnology (see Chapter 4). Tracts of natural vegetation and biodiverse pastures aside, these may be the most species-diverse structures in many landscapes (see Photo 11.1). As such, they fully mimic and can substitute for natural ecosystems. Their location within the landscape can be with this in mind, although use characteristics (e.g., DAPs) are more often the primary placement determinant.

These gardens are subdivided into forest, shrub, and homegardens. This is based on composition and productive emphasis, where the subdivisions roughly correlate with landscape placement. For example, homegardens are most often located near households, and forest gardens are generally found on marginal land distant from dwellings, while shrub gardens have a transitional placement, either as a temporal or spatial component, e.g., between seasonal staple or commercial crops and forested ecosystems.

PHOTO 11.1. This highly biodiverse and dense agroforest shows the density of the agroforest edge.

Méndez (2001) has suggested another use-based classification, again with placement implications. The subdivisions are

1. ornamental,
2. subsistence,
3. handicraft,
4. mixed production, and
5. minimal management.

The ornamental category is a garden, less as a source of food than to beautify the immediate landscape. These are often prominently displayed near dwellings. In contrast, subsistence gardens provide households with nonstaple foods, spices, and medicinal plants. On commercial farms, they are smaller and found near kitchens; on fully subsistence farms, they are larger and supplant or overlap the ornamental function.

Handicraft versions, a source of commercial raw materials, are situated near a center of activity (e.g., major road) on appropriate land. Mixed production systems yield a wide range of outputs. As these have both commercial and educational functions, they fall midway, both in use and placement, between the home and forest garden. The minimal management garden is equivalent to a forest garden where, exempting harvest inputs, little time and effort is invested.

## NONUSEFUL BIODIVERSITY

The type and form of vegetation in a landscape clearly determine the benefits received. The basic idea is that, through patterns (spatial and temporal), selection, and application of agroecological principles, human well-being can be maximized. More plants and plant species are considered as positive but, with appropriate caveats, this requires some planning and the avoidance of pitfalls. Weeds are one of these, as is the selection of trees or shrubs that can aggravate an environmental threat.

### Weeds

Unwanted plants are detrimental to the productive potential of any agroecosystem, and their removal can be an expensive labor input or require herbicide use. Although weed control is mostly outside the scope of landscape agroecology, it does have wider direct and indirect economic implications.

The basic control measures include the removal of all unwanted species, through manual or chemical (herbicide) applications. Biological controls exist where, through ecosystem design, i.e., shading, crowding, or the use of allelopathic plants, unwanted species are controlled.

Another strategy, partial weeding, retains some of the uninvited biodiversity with possible ecological gains (e.g., harboring predator insects). The methods vary: (1) an area of bare ground may be maintained around wanted species with only minimal intervention outside this zone (as with larger plants) or (2) only the more crop-competitive weeds are removed (where the less competitive weeds will help keep the more competitive in check, reducing the need for subsequent weeding). The latter may also be part of a fallow cycle where selective weeding leaves plants that are favorable to subsequent nutrient capture (Rouw, 1995). Fallow cycles can also serve as a restraint on future weed growth (Rouw, 1995).

Other localized weed control methods are more inclusive. Intense burning can destroy in-soil weed seeds; cover crops can smother new weed growth; decoy plants or plant residue can trigger species-specific parasitic weeds before the primary crop is in place, e.g., the maize weed striga in Africa (Rao and Gacheru, 1998). Landscape-wide measures are less certain and, although not documented, interactions do exist. It may not be beneficial to have areas of common and damaging weed species that can easily reseed cropping areas. They can be along plot margins, fallow areas, or on untended marginal lands. From this perspective, control has a landscape component.

## Detrimental Plantings

Individual plants may be beneficial, but large-scale planting, especially of exotic species, may be detrimental. For example, some plants are regarded as being thirsty. They soak up large amounts of water. When rainfall is abundant and well distributed, this is not a problem. In other situations, it can bring on negative consequences.

One such species is eucalyptus. Although it is a large species class and evidence is scant, some have reported severe moisture problems when these trees are planted extensively (e.g., Evans, 1992). The difficulty is that they take much more water than native forests and can dry streams prematurely.

The survival strategy of this species is best suited to areas with infrequent, but heavy rainfall. Here the ability of eucalyptus to appropriate large volumes of water may serve to reduce flooding. When rainfall is limited and more evenly distributed, large-scale plantings may diminish valuable runoff.

This does not mean that this species should not be planted. A few scattered trees do not overly affect area hydrology. Also, plantings in areas such as hilltops do not necessarily impact the overall landscape.

In addition to moisture, other detrimental plantings may affect wildlife or pose a fire hazard. It most cases, these plantings involve exotic species that are either planted as forestry species or have become a landscape weed. Highly combustible species should be avoided where fire danger looms large, or the planting should be positioned such that the danger is mitigated.

Wildlife is affected when a species does not provide favorable habitat and displaces those plants that do. In California, eucalyptus has been faulted both for flammable properties and a ruinous effect on a variety of bird and insect species (Williams, 2002).

## BASICS

A functioning natural ecosystem and equally potent complex agro-ecosystems derive their attributes from internal dynamics (nutrient flows, insect cycles, microclimate, etc.) that are jointly provided by all the flora and fauna present. This occurs when the dynamics generated by the sum of all component flora and fauna are greater than the effects caused by any one or two components. This is easier to define than measure. At some point, it is apparent that a functioning natural agroecosystem (one that duplicates the full spectrum of ecological dynamics found in natural ecosystems) exists within the boundaries of a productive system.

All flora-based natural ecosystems have similar baseline properties. How these properties are arrived at in an agroecological context is still an open question.

### Parameters

Biodiversity stems from three parameters: biological diversity, species density, and spatial disarray. This has been described as a $d_3$ (density, diversity, and disarray) ecosystem (Wojtkowski, 1998, p.104). The natural agroecosystems that follow presuppose that each of these three parameters exceeds some threshold value.

A functioning natural agroecosystem can conceivably tolerate less than full diversity, density, or disarray (see Photo 11.2). Reduction in one parameter may require compensation through an increase in another. The use of these parameters in specific agroecosystems is not fully understood and, across the landscape, less so.

### Diversity

With an ecosystem approach, the number of species needed is a subject of study. Approximately seven to ten species have been sug-

PHOTO 11.2. An agroecosystem showing biodiversity and disarray, while lacking density. This example, from Brazil, has cinnamon *(Cinnamonum zeylanicum)* as the primary crop.

gested as the minimum number needed to realize a functioning eco-system. This is reinforced through focused research and observation (e.g., Baskin, 1994) and is based on the premise that these species are niche-variable, with ample variation in the nonproductive DPCs for each species.

Most natural ecosystems far exceed this minimal number. This may be because more biodiversity can support the improved ecological performance of ecosystems (Naeem et al., 1994; Kareiva, 1994). This holds true in agroecology, where the numbers can top out at 40 to 200 individual species.

Another aspect to agrobiodiversity is the percentage of each species present. There is evidence that a functioning natural agroecosystem can exist even where there is a large dominant population of a few species. In some agroforest variations, up to 60 percent of the population can be one or two species. This may be accommodated within a natural context by (1) increasing the overall biodiversity well beyond a baseline of seven or so species and/or (2) ensuring that the populous and dominant species have sufficient niche variation to cover any existing niche gaps. Biodiversity also includes having different age categories for the species included.

Agrobiodiverse systems contain species of minor use and/or a comparatively large number of species classified as ornamentals. With biodiversity, if a plant or animal is exerting no obvious negative effect on overall productivity, it is regarded as having a positive impact.

The less noted forms of biodiversity are common weeds, in-soil and, equally undernoted, in-canopy microflora and microfauna. The fauna component enhances soil fertility, moisture absorption, and a host of other ecological roles. A common example is the earthworm, which promotes a range of ecological functions and thrives in permanent and diverse ecosystems (Hulugalle and Ezumah, 1991). Aboveground, ants can accomplish similar objectives (Stanton and Young, 1999; Risch and Carroll, 1982).

*Disarray*

With disarray, there is less evidence to support any broad conclusions or suppositions, and the amount of disarray required is an open question. The key aim with biodisarray may be having a variety of individual interspecies interactions, and disarray promotes this. These can be above and belowground.

Biodiverse, ordered agroecosystems exist, and with sufficient density, interspecies contact is fostered and the desired outcome is achieved. With ordered systems, there is a tendency to avoid crowding and to refrain from uneven ages within groups of like species. The effect is to proportionally weaken existing or potential natural dynamics.

With a disarrayed or ordered formulation, a visual clue that resource and ecological inefficiencies exist is where sunlight strikes bare ground rather than leaves. Full disarray with competing plants will evolve into a multistoried system, and this encourages a wide range of efficiency-promoting dynamics.

This does not imply that with disarray patterns cannot exist, only that those which do supersede those commonly associated with agriculture. The two basic patterns, minimum interface and midpoint, put the emphasis on understory or overstory species (Wojtkowski, 1998, p. 87). Across the larger landscape, these are illustrated in Figure 9.3.

The emphasis for the midpoint design is, by allowing taller species to capture more sunlight, overstory production (see Figure 11.1). To achieve the best LER, the shortest plants are shade tolerant, can be paired with taller species having as a DPC reduced canopy spread, or have the primary species in the upper canopy levels (as with a wood-producing agroforest).

In contrast, resources can be more favorably allocated with density through a subminimal version of a minimum interface design. If the understory around a taller plant starts with sufficient light and there is

FIGURE 11.1. A disarrayed midpoint design, in which plants are located midway between two taller species.

sufficient density and niche variation to deny the taller species full access to belowground resources, the net effect is to emphasize more the productivity from the shorter-statured species. The type of close spacing used is shown in Figure 11.2.

The minimum interface version (without subminimal intent) does not rely on belowground competition to direct essential resources to certain species. Instead, it is only a means to allocate light to the various levels. For this, the ordering remains the same (tall to short), but spacing is more open with less interspecies competition. Photo 2.1 shows a minimum-interface ordered row pattern.

There are documented cases where these two patterns are combined. In these cases, for a few overstory species, there exists a widely spaced midpoint while the lower levels are placed directly adjacent to the overstory species in minimal interface design (e.g., Jensen, 1993).

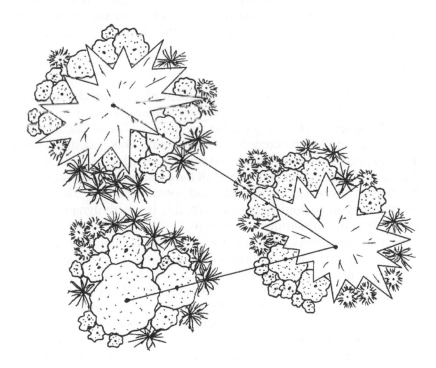

FIGURE 11.2. A pure disarrayed minimum interface layout in which each species is located next to a taller species.

*Density*

As with disarray, the parameters of biodensity are less well known, but they have some flexibility as long as the end goals are kept in mind. With density and close association, plant-plant interactions are implied. Density includes the interstem distance and the vertical and horizontal intercanopy distances.

In order to maintain a constant number of plants per area, managed biodensity can require, through pruning, keeping plants in a short-statured or narrow-canopied juvenile state. This can reduce the per plant output but, because of the high biodensity, the total per area output can be high. The sustainability issues that arise are countered, e.g., through increased density, more nutrient capture, or the non-harvest option.

A functioning ecosystem can exist in weakened form where density is less than its fullest. When this occurs, ecological inefficiencies impede growth (as with the inefficient use of light) and some of the subsidiary mechanisms (e.g., microflora and fauna) begin to thin.

### Landscape-Wide

At the landscape level, a naturally functioning agroecosystem can exist, but not always in the form established using dense, diverse, and disarrayed systems. On a larger scale, natural ecosystems can be a mix of dissimilar and less biocomplex smaller units. This is demonstrated in natural grasslands, where trees and small forested areas are scattered throughout the landscape. For these, all the component ecosystems exhibit density, diversity, and disarray and contain basically similar dynamics. From an ecological perspective, this is where dynamics flow seamlessly across dissimilar ecosystems.

This may not always be the case with agroecological landscapes, in terms of both ecological qualities of component agroecosystems and dynamics contained. Depending on the landscape design, weakening of density, diversity, and disarray may affect the strengths of various natural functions, where some landscapes may contain only segments of what can be found in a fully functioning natural ecosystem.

For example, strong natural predator-prey dynamics may be lacking, but, to the good, insect movement may be restricted. This may be a case of maintaining the SIZ for a specific interaction, in this case a repellent effect. If not highly visible, these dynamics may go unnoticed and, if enough of them are lost, one or two individual species begin to dominate and the advantages of having a natural agroecosystem dissipate.


## MANAGEMENT

A highly biodiverse landscape can require some changes in normal management patterns, which can challenge many cultural norms that have evolved from ecologically simplified and more ordered landscapes.

One large cultural jump is biodisarray, where the use of unordered agroecosystems may run counter to a society's desire to introduce order. However, it is a component of a fully functioning ecosystem where any deviation can be met with a reduction in intended DAPs. Also, the nonharvest option, with its apparent waste of potential output, has purpose, that of recycling elemental nutrients for subsequent growth.

Other changes involve how systems are managed through the inputs used. The basic unit of management in many agrotechnological landscapes is the plot. In a biodiverse landscape, this may not always be true. The concept of the plot loses integrity when the contained ecosystem must also take into consideration what transpires outside of plot bounds or where the needs of individual, ecologically contributing species must take precedence.

Without independent plots, other management techniques come to the fore. Among these are rule of thumb and community net present value.

The difference often lies in how the landscape is viewed culturally (e.g., individual plant, plot, agroecosystem, or landscape based), rather than the outward visual appearance of the different ecosystems. The difference can be demonstrated with a parkland agrotechnology, where the system is viewed as either a set preplanned plot or two subunits, trees with crops and crops with trees.

These techniques apply to both agrobiodiverse farm and forestry silvicultural situations. As mentioned, foresters have developed silvicultural methods for achieving the desired results. Subsistence farmers also have developed techniques for overseeing biorich ecosystems, which follow along the lines of those presented here.

## Rule of Thumb

In large part, the effective application and management of bio-diverse landscapes depend on an accumulated knowledge of local species and growing conditions. Where extensive knowledge is lacking, there are some overall rule-of-thumb guidelines that can help manage a productive $d_3$ environment (Wojtkowski, 1993).

In complex landscapes, the focus is on individual plants, and these rules apply to a biodiverse landscape managed without agrotechnology as the basic unit. Modified, the rules are as follows:

1. If the output from a plant or group of plants is needed and producing well, leave it; if not, alter the competitive environment.
2. If its production is not needed, neglect it.
3. If it is negatively affecting a more desirable output, prune it.
4. If space exists and essential resources are unused as measured by the amount of light striking the bare ground, plant or let something grow.

With this management plan, a fairly efficient system, in terms of desirable outputs, can evolve as long as a $d_3$ landscape is maintained. This management plan does allow considerable design flexibility, enabling land users to maintain the desired patterns, e.g., minimum interface, midpoint design, or some mixed pattern, while achieving high ecological efficiency.

It must be remembered that to maintain the full advantages of biodiversity, species are only eliminated as a last resort. As an example, Vieyra-Odilon and Vibrams (2001) found 74 varieties of valued weeds in maize fields, including forage, potherbs, medicinal, and ornamental plants. In Côte d'Ivoire, of the 27 wild tree species used as shade above cocoa, 13 provide firewood and medicine, 11 provide food products, and 6 are used in construction (Rice and Greenburg, 2000).

The tenet behind biodiversity-based management is not to eliminate a plant, but to reduce competition. Weeding is temporally selective, occurring, as in the maize example, only when the maize is most vulnerable. Other options include spatial weeding (nearby weeds only), niche selective weeding (certain species), and/or, for larger plants, pruning.

## *Community Net Present Value*

As an alternative or supplement to rule-of-thumb management, there is community NPV management. With this method, the landscape is subdivided into overlapping subunits, which can be individual plants and/or groups of species.

The land user assigns a value to each subunit based on the NPV of the unit and a discount rate as perceived by the land user. The objective is to maximize the NPV of as many of the subunits as possible. The NPV approach takes into consideration future as well as present outputs.

In this management system, each plant has negative or positive values in the assigned subunits. Positively valued plants are more intensely managed. Positive influences include facilitative effects, present uses, and NPV of any future outputs (wood, fruit, etc.); negative values include excess competition affecting other plants. When a species negatively affects a number of subunits, the plant is ignored, pruned, or, if it is an extreme negative influence, removed. If the essential resources available to a group seem underutilized, more species may be added. This management method may seem complicated, but in use and with an experienced practitioner, superior results are achievable commensurate with the level of user knowledge.

## *BIODIVERSITY RANKING*

In planned and managed landscapes, enhanced biodiversity is the notion that groups of specifically selected plants can accomplish set environmental objectives. This is the idea behind intercropping and complementarity, where two selected species can coexist or thrive with corresponding gains in useful output. It extends to complex polycultures, such as agroforests, where a mutually supporting designed and managed ecosystem can benefit all plant components.

Landscapes can be evaluated on how they employ biodiversity and the associated ecological dynamics that result. Four ranking categories are possible: (1) coverage, (2) intensity, (3) eloquence, and (4) holism. Each defines a different perspective, and the most ennobled, holism, has the best cumulative mix of all. Inherent in all these is the concept of optimization.

## Coverage

The most basic measure, coverage, is the amount of interspersed biodiversity, e.g., number of different species and the amount of interspecies interface in the productive landscape. This can be achieved through the placement of single or multispecies systems. The objectives are met by combining inter- with intraplot effects through small plots and large amounts of interface. The other option is to focus less on the spatial options, opting instead for a more temporal approach.

Ecological gains follow from having a varied landscape of mono-cultural, fallowed, and/or mixed-species systems with smaller plots, where the individual ecosystems can be placed to better champion desired effects. This might confer better insect predator-prey dynamics. Using a windbreak example, the use of a single species within a larger landscape shelterbelt system with an effective wind-protecting SIZ contributes to coverage.

## Intensity

Beyond coverage, there is intensity, where the role of each plant and how intensely it is used is included. Intensity carries with it an economic component where broad, multispecies, plant-plant interactions within the landscape bring about environmental and economic benefits. More intensity equates with greater ecological efficiencies and agroecological potential. With intensity, there is more intent in the overall design than with coverage.

Plant-plant interactions are paramount in any agroecosystem. These also occur through intersystem complementarity (favorable intersystem DAPs), transitional interfaces, or use of buffer species. Plant-plant dynamics that span time periods are also part of this, as well as the dynamics that follow from having zones of natural vegetation.

With intensity, a windbreak should accomplish a range of tasks. Windbreaks are composed of multiple species, each having complementarity with neighboring crops through design (e.g., buffer species) and each contributing, in small or large measure, to the overall ecology of the landscape.

## Eloquence

In contrast to intensity, eloquence refers to effectiveness, vitality, and sophistication in both biodiversity selection and plant-plant interfaces. This is a step beyond intensity, looking at species selection and the various levels of interaction. These can be temporal and spatial and combine perennial and seasonal species.

Eloquence is a measure of how well plants are used in a broader context, how well a site is matched with crops and agrotechnology, the use of auxiliary systems with regard to the primary ecosystems and topography, and the effective use of a temporal flow. There are many more planned interactions, which are achieved through diversity and placement of vegetation.

An example of spatial eloquence is hillside species and placement. The agroecosystem can be different in terms of species and possibly density at different elevations. With eloquence, the natural dynamics are more broad-based, less specialized, than with a holistic design.

An eloquent windbreak has all the previous characteristics (coverage and intensity), but offers more than just low-level influences. Planning may have added key predator-prey dynamics and a nutrient capture role. The windbreak may be positioned to serve as a waterbreak or riparian buffer.

## Holism

The last measure, holism, is of a landscape that achieves the set goals, both environmental and productive, through mix of ecosystems, plot size, individual placements, and timing. Ecosystems exist that promote those ecological activities and contribute to a productive outcome. This can involve the intense association of individual and community species in overall plant wealth (Jain, 2000).

Here both individual species (plant-on-plant dynamics) and functioning ecosystems encourage ecological dynamics. Similarly, the flora and fauna that are detrimental to productive purposes are discouraged, but not necessarily eliminated, through management and ecosystem design.

Carrying on with the windbreak example, any number of species can be utilized to accomplish specific tasks. The species mix and structure can encourage favorable insect predator-prey dynamics in the accompanying SIZ. Crop-eating birds may be discouraged by tree species that are not conducive to their presence or that are conducive to other animals that discourage the birds. These functions are in addition to what is gained from coverage, intensity, and eloquence. This windbreak is ecologically and economically interlocked in time and space with the principal-mode systems and reinforcing the dynamics (both temporal and spatial) inherent in the productive systems.

# Chapter 12

# Other Topics

There are topics that, on the surface, have little to do with ecology. They are part of the human-nature interface and come into play through their social, economic, and productive influences.

## HOUSEHOLD LOCATIONAL PATTERNS

Most farming landscapes contain the households of the participants, where location is partially a cultural norm and partially dependent on cropping needs and topography. In some respects, this is less an agroecology topic and more sociological. At the least, resident location provides limits to the amount of interplot agroecology that can be consolidated in the broader landscape. At most, it can, through transport, ownership patterns, and soil fertility needs, dictate the type of agriculture and forestry practiced. The four types of relative dwelling placements are (1) strong village, (2) scattered permanent, (3) scattered temporary, and (4) nomadic. Except for the last, these are illustrated in Figure 12.1.

### Strong Village

One locational norm is a strong village approach. There are a number of variations. The dwellings are around a village core (usually containing shops, a religious structure, and maybe a government building). Farming and forestry activities take place in various rings that surround the village core. The standard layout has, outside the houses and farm buildings, a circle of gardens, then orchards, fields of staple crops, pastures in the outer agricultural ring, and furthest from the village center are forestry activities of varying intensity.

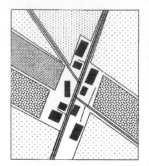 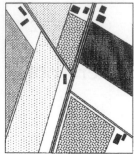 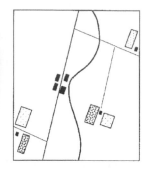

FIGURE 12.1. Three different dwelling patterns. From left to right, these are (1) strong village with surrounding fields, (2) scattered permanent where the dwellings are located away from the village core, and (3) scattered seasonal where people live away from the village only during the cropping season.

This type of pattern has been used in Europe and can still be observed (Photo 12.1).

As the village grows, this pattern dissipates. Some intensive agricultural activity may remain within the confines of a town while less intensive and larger areas are at the periphery.

Another variation has people living in villages, where the fields and forests are scattered about. This may be the result of land ownership patterns. Locationally, more intensive activities are closer to the villages with less intensive activities further away. The problem with this is that, with scattered rights and ownership, a coherent agroecological landscape design may be difficult to achieve.

### *Scattered Permanent*

In contrast to a village core pattern, farms and farm dwellings may be scattered across the countryside, where each household is separated from each other and any hamlet or village (see Photo 12.2). The Midwestern United States exhibits this type of pattern where agroecosystem locational needs are set by the placement of households, farm buildings, and roads. In the absence of cooperative agreements between the entities, this may offer the best opportunity for interplot agroecology.

PHOTO 12.1. The strong village model. This picture looks inward from the garden ring, past the barns toward the village core with the church steeple. This photo was taken in former East Germany.

PHOTO 12.2. A permanent farm household that is far removed from any village center. This photo is from the Chilean Andes.

## Scattered Seasonal

There are advantages to living in villages (social activities, work opportunities, etc.) while each year there are the same pressing agricultural needs. Therefore, it may be expedient to live within a population center for part of the year and move into distant fields while crops are being raised. Long travel distances can be a function of land ownership and/or the lack of suitable ground coupled with the need for a continual presence.

This type of habitation pattern is found in regions of Africa, where wildlife can decimate crops if humans are absent (even a few hours) and/or where lengthy slash-and-burn rotations can put agricultural fields far from villages. In New Guinea, this seasonal movement is observed where the rugged terrain prohibits ease of travel and limits the area that can be cultivated. In these cases, temporary housing can be quickly erected or dwellings easily repaired with local materials.

## Nomadic

People may move constantly in response to climatic need and in areas devoid of obvious legal boundaries. The participants usually have portable housing (e.g., tents) and can move with minimal effort. There may be a sequence with some cropping in bottomlands during a brief wet period, with the remainder of the year spent chasing unexploited grazing land.

# TRANSPORT (ACCESS)

Roads, rivers, canals, and railroads are not only local landscape features, but serve as a two-way conduit to markets. They can have a profound influence on crops and cultivation methods. As an example, the loss of a railroad in Central America caused farmers to shift from export bananas to subsistence farming (Soluri, 2001).

Transport problems can force a change to crops with value-added potential. This is where, through labor inputs, locals convert low-value bulky crops to a higher-value, more portable, deliverable, and saleable state. Mexican farmers in remote mountains, through stills and distillation, convert low-value sugarcane into higher-valued, and illegal, alcoholic spirits for sale outside the area. More conventional cases include wood converted to charcoal, straw into brooms, and fresh fruit into a dried form.

# ROADS

Roads and paths connect the different areas in farms, plantations, and rural landscapes and, depending on quality, determine land use through movement of labor and inputs and removal of outputs. As such, they are a necessary part of agroecology.

In farming, they can converge at dwellings or farm structures and, as such, be in line with housing placement or delivery requirements. In forestry, roads may centralize upon log landings and sawmills. Whatever the case, well-thought-out placement can improve operational efficiency, and roads can be part of the agroecological landscape.

## Types

The three types of landscape roads are (1) haul, (2) access, and (3) strip (FAO, 1976). The first of these, the haul roads, is subdivided into primary, secondary, and feeder (Figure 12.2).

The primary haul roads have the heaviest use and, because of this, are of the best quality. In most landscapes, land users try, wherever possible, to use public or government roads as the primary haul roads.

The secondary roads connect groups of plots with primary roads while the third category provides efficient passage between fields and/or farm or forest structures. On subsistence farms, the third type can be paths and, on more commercial enterprises, they may offer an efficient route for machinery, but not to haul inputs and harvests.

Strip roads (Figure 12.2) serve two purposes, hauling and access. They are found in forestry or farm forestry, finding use for infrequent tree harvests and, between the sporadic wood harvests, serve as entry to forests for alternate products and/or hunting.

The temporary feeder or skid roads, although mostly confined to forestry, are also found in some plantation situations. In forestry, they are almost always abandoned as functioning roads after a harvest, but may continue on as forest paths (Figure 12.2).

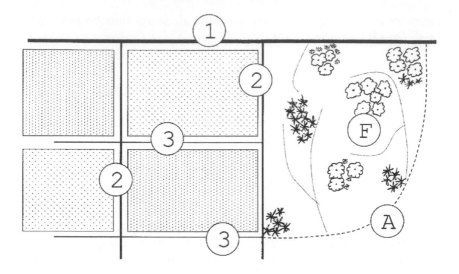

FIGURE 12.2. The types of roads found in rural environments are haul, strip (A), and feeder (F), with the haul roads being divided into three categories (1, 2, and 3).

## *Placement*

As conduits, roads must accomplish their function without taking unreasonable amounts of prime land from productive use while, at the same time, providing the quickest and easiest means of access. For productive efficiency, roads are best put on more marginal lands with the proviso that the route is efficient (most direct). Straighter roads reduce construction, maintenance, and hauling costs while usurping less overall land area. Compromises are the norm with road placement, for example, using the base of a hill for a secondary road to avoid using prime bottomland makes the road longer.

The basic idea is to transport outputs by truck rather than more expensive tractors or, in some regions, animal conveyances. Trucks require better roads or firmer soils (primary and secondary haul roads) and, to this end, the route should be planned according to use. For example, oil palms *(Elaeis guineensis)* are highly productive throughout the year. Given the sheer volume of output (up to 20 t/ha per year), an integrated road system is advantageous with numerous feeder roads connecting to the secondary or primary haul roads.

## Agroecological Uses

Roads offer few opportunities for vegetative growth, but can serve as barriers such as firebreaks, buffer zones, etc. In this double purpose (transport and barrier), placement is key. Ridge tops are a prime location, as they offer a flatter surface, better drainage, haul efficiency, and a good firebreak position.

A road may act as a buffer zone. If a row of trees is highly competitive with an adjoining crop, a road located in between serves as an interecosystem buffer.

Roads are not always totally nonproductive. In the oil palm example, the feeder roads can be overtopped by palm canopies and, with the possible exception of soil compaction, this has little effect on tree output and spacing. Intercropping possibilities are diminished, but only in every other row.

## GENDER CONCERNS

For many farming and forestry activities, cultures assign tasks based on gender. Commonly, duties assigned to males include land clearing for crop planting and harvesting of staple crops. Females may undertake gardening, planting chores, and firewood and water collection. Male-dominated activities can be more expansive in terms of area covered. Because of this, locational needs are recognized as requiring, and often having, a gender input.

## LANDSCAPE FUNCTIONS

Some basic landscape functions can be usurped by other needs. One of these, the saving function, may not always endure where sound banks and other monetary institutions exist. The second of these, the educational cycle, is where land users explore the DPCs of new species, varieties, and cultivars. An educational cycle is less needed where experimental stations are actively engaged in appropriate field trials.

## Savings

In many cultures and regions, land users need to generate or save money for educational, marriage (e.g., dowry), or other expenses associated with offspring. Some cultures save money through livestock; others use forestry tree species. In both cases, herd or tree growth, they provide a marketable commodity when cash is needed.

In Kenya, the tree species *Melia volkensii* serves a savings function when, upon the birth of a child, the tree is planted in anticipation of future school fees (Kidundo, 1997). The presence of savings institutions does negate the savings function. Pines are used in the southern United States to cover future college or marriage costs.

In contrast to trees, cattle may be a less effective means of savings. In addition to more risk, this can be a zero sum game where, above certain stocking rates and through overgrazing, there are losers in both contained and free-ranging situations (Bradburd, 1982).

## Education

With some cultures and regions, the introduction of a new, often exotic species presents a dilemma. People may recognize a plant's value, but do not always know the establishment, propagation, and growth needs, the quality, desirable agronomic properties, market potential, and/or use properties (Styger et al., 1999).

Many of the practical questions must be answered through trial and error. For grain and other seasonal staple crops, the optimal spacing is a useful piece of information, one that can only be determined in the field.

For other plants, mainly perennials, a multitude of properties (DPCs) need to be ascertained, including the growth rate, use of essential resources, soil requirements, moisture needs, and other site and climatic questions that must be answered before the plant is part of a progressive and efficient landscape. On-site determination of these questions is part of educational cycles, and traditional land users do reserve portions of plots for experimentation (Johnson, 1972).

In some cultures, complex agroecosystems (e.g., forest gardens) are the ideal place to study the properties of perennial plants (e.g., Smith, 1996). They contribute to biodiversity without substantially interfering with the growth and productivity of established species. Lacking a complex agroecosystem, these plants may be located at the fringe of a household area, between gardens and fields, where they can be observed and managed.

## *Domestication*

Part of domestication (bringing a plant from a wild state to one where it better serves human needs) is part of the process of education. In tropical regions, there is no lack of edible flora or species that produce useful nonfood products. Many of these are still found in natural rather than planned or managed environments (e.g., Aman, 1998; Mertz, 1998; Styger et al., 1999; NRC, 1996b).

There are some recognized goals in domestification (modified from Gerritsma and Wessel, 1997):

1. In the first step, the plant is studied, often informally, to find the best varieties, growing environment, and any constraints on propagation and yield.
2. A second step is to maximize yields (e.g., a better harvest index), improve fruit (or other output) quality, and eliminate any undesirable traits (e.g., thorns). This is an ongoing process where selection, cross-breeding, and/or improved management brings to the fore the desirable properties (e.g., Lovett and Naq, 2000).
3. Another objective is to expand the range of sites in which the plant will grow. Again, some plants can be utilized directly from their natural state; others may require selection and breeding before widespread use occurs.

How and where this occurs has a placement component, where land users test new varieties in appropriate venues. Once usefulness is recognized, land clearing, leaving in situ the desired species, can be the first serious look.

Work begins in earnest when a species is planted or moved to a semiwild environment (e.g., an agroforest) or other less-managed stand. Once some understanding is gained, then more integration into the mainstream economic sphere is in order (Gerritsma and Wessel, 1997).

Placement is important in this process. If land is lacking, the educational and domestification opportunities may not exist (den Biggelaar, 1996, p. 71). One alternative is to use new species in existing roles that are not critical to the productive capacity of an enterprise. For example, antierosion species may be gradually replaced or augmented by new species that are perceived to have suitable properties. For productive species, room may be made near households or at the edge of active stands so observation and active management can be better undertaken.

# WILDLIFE

In Chapter 8, the ecological uses of fauna (bird, bats, and small mammals) to control various pests are discussed. This does not exhaust the topic of wildlife management, as natural fauna can be a source of diversion, sport, and/or subsistence. Wildlife includes insects (such as butterflies), birds, fish, small and large mammals. There are differing views and options on the role of wildlife in the landscape, ranging from (1) outright preservation and noninterference to (2) conservation to maintain viable populations to (3) semicontrolled husbandry coupled with hunting.

## Preservation

Outright preservation takes two forms: (1) selected species and (2) full ecosystem. Selected species is less common and involves prohibitions on killing specific endangered species. This may go as far as forbidding the destruction of key habitat, e.g., the nesting grounds for a bird species. It should be noted this may also affect plant species. For example, many governments forbid the hunting of certain animals or the cutting of protected plant species. The disadvantage of this strategy, with both animal and plants, is in enforcement.

The most visible efforts at wildlife preservation involve large-scale land set-asides in the form of parks and protected natural areas. Many of these strive to keep natural vegetation and associated fauna while permitting some human activity, e.g., recreation or forestry. Protected natural areas seek to minimize human impact, allowing the existing ecosystem to remain as undisturbed as possible or feasible.

A lengthy discussion on undisturbed areas is outside agroecology. However, there are issues that overlap into the agroecological landscape. All involve encroachment, either of (1) local peoples into set aside areas or (2) wildlife into farm communities. The idea of a totally protected area may be acceptable in some countries but, in others, neighboring peoples may regard forest resources as a birthright.

Many recognize that limited use of protected areas can be better than prohibiting access (McNeely, 1993). By controlling what is taken, total destruction is avoided, while economic incentives are provided to maintain those vital functions that need protection.

Another issue is wildlife movement into populated zones. Again, there are opportunities for cooperative arrangements where more common animals are hunted outside the protected area, while the endangered species are spared (Nepal and Weber, 1995). This can be part of a managed hunting strategy (discussed in a subsequent section).

## *Conservation*

The preservation of natural wildlife with an agroecological landscape is often a regional endeavor, one that need not always run counter to productive purposes. Some species can exist within a small but favorable habitat, and others require a larger area, while a few experience difficulty with human encroachment and the loss of undisturbed territory.

There are baseline high and low figures that can help in determining area needed and territory viability (Charles, 2002). Below the low figure, there is not enough favorable habitat to support a breeding population. The upper figure is equally of interest, as above this number there is a decrease in per hectare and per species benefits. This does not mean that added area in excess of the high figure should not be contemplated, only that there is more leeway for any productive-conservation compromise above the point where a viable population is ensured.

Studies have approximated high and low values of 15 percent and 5 percent of the total land area (Charles, 2002). These values may be subject to modification, depending on species, individual habitat needs, and other factors.

## *Habitat Retention*

Taking into account these values, the opportunities for conservation are many and varied, and the areas set aside for some wildlife need not be extensive, only well-placed. For others, e.g., forest-breeding species, fragmentation in an agricultural landscape can be quite detrimental (e.g., Donovan and Flather, 2002).

For hunting, larger areas seem best and, if coordinated with neighboring holdings, less than optimal sections of natural vegetation or other wildlife-favorable areas can be made spacious enough to support a wider array of wildlife species, more so through intensive regional or interfarm planning. Putting small, unused areas at the periphery of farms, such that they directly border similar areas on other holdings, can create larger areas.

## Corridors

Natural corridors within a landscape, connecting patches of natural vegetation or favorable habitats, are also part of placement. They allow fauna to traverse wider distance with less danger (e.g., tree-dwelling species) and/or to conceal wide-ranging but shy animals. These strategically placed links can effectively extend smaller patches of vegetation. Corridors can also be riparian buffers or located along hillsides or ridges where they can double as watershed or antierosion zones.

## Enrichment

Another positive fauna measure is system enrichment, not in productive species, but in those species that encourage certain types of wildlife. This can involve multipurpose secondary species attractive to natural fauna. They need not detract from the primary purpose of a system, as wildlife can be part of an agroecological plan (as with insect-eating birds) or a minor annoyance that does not overly concern land users.

Cooper et al. (1996) document an example from Java where, of 121 bird species found in agroforests, 15 are endangered. Perfecto et al. (1996) reached similar conclusions with birds and shade systems in Latin America. Agroecosystem mimicry or other encouragement measures may have the potential to reduce the amount of area a given species needs to survive or thrive. As with any fauna management topic, brief discussion can only suggest broad approaches. The promotion of a particular wildlife species may require specific measures.

## *Hunting*

Conservation can have another more practical purpose, maintaining wildlife not as part of a conservation ethic, but as a continual food supply. The role of the landscape in providing natural game for sport or subsistence is well established. There are cases where the value of game exceeds the potential for raising domestic animals. As an example, hunting rights in the commercial pine plantations of the southeast United States surpass in value any grazing rights. In this case, deer are encouraged, but little allowance is made in the way of landscape design for their propagation. The value of the pines exceeds the gains from leaving open spaces to promote optimal deer habitat. The cost of fencing, along with the management requirements of grazing, are major components in this equation.

This idea can be carried further. Natural game can be a substitute for domestic animals (Linares, 1976). Besides eliminating the work, costs and risk associated with farm animals, natural game requires no dedicated farm infrastructure (barns, pastures, fencing, etc.). A well-designed landscape attracts animals, removing the need for extended hunting trips. The cost, in terms of crops lost, is looked at as a tradeoff between carbohydrates expended and protein gained.

The type of animal suited to this approach is not too small nor too large, and is comparatively easy to hunt or trap. It is not capable of rapidly destroying large areas of cropland or breeding out of control, and can be discouraged from entering certain areas.

To accomplish this unconventional strategy, some modifications must be made in the landscape. Redford et al. (1992) found certain trees species associated with common game animals, many of which are wild or domestic fruiting trees. Such trees can be a favorable off-shoot of the nonharvest option in enriched forest ecosystems, where only the better quality or higher valued fruit is removed. Peterson (1981) has suggested more forest edge and broken cover with areas surrounding crop plots. This is best accomplished through a habitat formulated to encourage the desired species, crop planting designed to absorb loss, trap crops (such as forest fruit trees), and sustained yields to maintain a base population of the desired animals.

Also needed is sufficient land area for game habitat. Enriched forests or agroforests can play a vital role in attracting and maintaining game populations.

## *SOCIAL FORESTRY (GATHERING)*

The involvement of a local population with intense forestry often involves taungyas or farm forestry where trees and crops are mixed. Social forestry can also entail the interaction of local people in forestry holdings (e.g., large-scale plantations) or protected forests. In keeping with the idea that controlled use is better than pure protection, this has broad ramifications that include hunting (as mentioned in the previous section), harvesting, or gathering. It is the last two topics that are addressed in this section.

Harvesting of timber in protected areas is commonplace and, if done in accordance with established principles, has less impact on the ecology of an area. The gains for locals come through employment or profit. The need for a well-regulated, sustainable process, at times not observed, does not require comment.

Logging and sawmilling activities are not the only realizable gains, and gathering can take many forms. Firewood is commonly acquired and, if done with guidance and in accordance with a management plan, can be a positive addition. The firewood obtained through tree thinning and/or pruning will both increase the value of standing timber and provide locals with wood.

If demand is high, foresters have the option to plant a fast-growing firewood species in plantations or natural forests. They serve as guide trees, improving the form while encouraging ecological interactions that can speed the growth of the primary and protected species. These plants are removed before the canopy of the primary species closes.

Other forms of gathering are possible; including mushrooms, berries, medicinal plants, fruits, and a host of such items. These can be overpicked, but if harvests are kept below sustainability limits, normal, well-managed forestry activities will generally not interfere with gathering. For example, Kendell (1980) showed that the gathering of mushrooms and berries in Sweden is not impaired by logging and other forestry activities.

## *LAND TENURE*

Most of this text has assumed that land users exercise full land control. In the previous sections, the topic of protected areas with access is briefly touched upon. Clearly, the type and form of land tenure have powerful influence on land usage.

Where land users have only a short-term lease, there is little impetus to make any lasting investments and improvements. The worst situation is where there is no ownership or control mechanism (e.g., unregulated common lands).

## Ownership

Clear legal title including all land rights is the most desirable situation. Then the owner has the ability to establish long-term agricultural and forestry systems. Despite ownership, there are limitations of control in the form of legal infringements, zoning restrictions, cultural norms, and other encroachments.

Culture can play a part in the degree of control exerted. It may be customary for owners of large areas to permit minor forms of hunting and gathering to take place. In other regions, traditional practices may be more invasive and reduce the landscape design options. For example, in parts of West Africa, if an individual plants a fruit tree on the land of another, that individual has the right to harvest the fruit.

Physical ownership patterns can contribute to numerous inefficiencies. Land users seldom have control over enough topography to produce an ecologically balanced economic entity. One land user may utilize bottomland, while another has hillside sites. Both may require the same mix of outputs, but prime land for a hillside farm might be a very marginal site for a farm located on prime bottomland. Inefficiencies occur when one land user plants crops on erosion-prone hillsides, while another land user has low-value tree plantations on prime bottomland. Similar inefficiencies occur when water control measures cannot be fully implemented due to the patterns of land ownership and the lack of cooperative agreement.

## Cooperative Agreements

In order to achieve goals, landowners can band together for a common purpose. Although rare, agreements can involve environmental or agroecological issues. This is more prevalent where the benefits and/or threats are real, immediate, a solution is within easy reach, and all will gain. Irrigation systems, with their clear benefits, require a high degree of leadership or control to fully implement and can require an enforcement mechanism to maintain an equitable outcome.

These agreements seldom extend to less apparent, more marginal situations. Landscapes and land users that can reap agroecological benefits will, more often than not, miss key opportunities due to poor land division, lack of agroecological input, and inability to find common objectives and/or reach a cooperative arrangement. There is some evidence to suggest that such agreements are feasible and more within reach than practice suggests (e.g., Jensens, 1990).

### The Community Landscape

In the previous section, a number of landowners banding together can improve agroecological gains. Other landscape-wide situations can be described where legal boundaries pale next to economic necessity.

One of these is the need to extend the productive landscape beyond the boundaries of a legal entity. One example may be a primary process industry, e.g., sugar, fiber (including wood), or other enterprise, that requires a raw material source, and a landscape where a number of entities must work in unison to supply these needs.

#### Multientity

Multientity activities span various holdings. A large company may contract with landowners to provide raw materials. This is common in agriculture and forestry where, because of manufacturing concerns, the need for a continuous flow of raw materials may be an agroecological force on a regional basis.

There are other multientity influences. Mechanized plantings and harvests may require specialized equipment, which is more economically used if put to greater use. It is not uncommon to share or rent such equipment. There are other cultural or convenience-based arrangements that spread labor and equipment across legal entities.

Although the agroecological impacts of such arrangements may be small, the potential exists for productive gains or losses. In a previously mentioned example from West Africa, planting the rice fields of village chiefs is a community activity. This is not entirely an altruistic arrangement, as the first fields to mature are subject to the severest attack by rice-eating birds. The others, subsequently planted, enjoy less predation.

## Multiparticipant

Multiparticipant situations are best served by having more than one person or group utilizing the same plots. Commonly, these are taungya based, where foresters manage the trees and farmers tend the understory crops. Because of the long-term nature of the resulting forest or tree crop plantation, control of the land usually rests with the foresters. Although the exact relationship is negotiable, the farmers generally provide a service in land preparation and weed control and benefit through the crop planted.

These arrangements exist in clearly defined circumstances. Unless the roles of the participants are well enumerated, the gains can be negated. In another case from West Africa involving a crop-tree taungya, the farmers planted the crops too close to the newly planted trees and the crop shading retarded tree growth.

These cooperative ventures can be more widely employed if an equitable multiparticipant agreement is achieved. Compromises in agroecosystem design can accommodate and offer more to each participant if studied in detail (e.g., Wojtkowski et al., 1988).

At the landscape level, multiparticipant arrangements both permit and require a large area to function on a continual basis. As a result, they will alter the community and land use situation over a wide area.

## Commons

One of the most difficult situations is where land ownership does not exist or where there is community use without control. The tendency is for all to take and few to replenish. This situation is referred to as "the tragedy of the commons" and, as the name suggests, the end result can be overexploited, badly degraded land with little productive potential.

This is not always the case. Where land pressures allow for a recovery period or less use intensity, the threat may be delayed. Protection can be provided by cultural, historical, political, socioeconomic, and ecological traditions (Tucker, 1999) or through education on the value of the resource (Chandrashekara and Sankar, 1998). An example is holy ground with an informal protected status.

## POLICY

Direct or indirect government policy has profound implications at the landscape level. These are the land allocation issues raised in the previous sections. There are others, tax and agricultural support programs, that dominate policy and set the agricultural agenda. This can have profound implications on the types of land use practiced.

One aspect that works against biodiversity is the fixation on single yields. Because they are easier to assess than complex intercrops, this can bias policy toward monocultures (Godoy and Bennett, 1991) and run counter to biodiversity advantages. The fixation on single crops may be carried further, by requiring direct action against insect and disease pests, rather than slower-acting and more subtle ecological controls. A simpler landscape design, one devoid of ecological frills such as auxiliary and mixed-species systems, also finds favor in the government policy arena. The need for clarity underlies price support and risk insurance and strongly biases the landscape toward a basic, uncomplex, less agroecological form.

For governments, there can be a strong inclination to have few commercial farms rather than many subsistence farms, as subsistence farmers pay proportionally less in taxes. Politicians may have other agendas—some may gain support through farm collectivization; others may want land distribution. Aside from these larger and often unstated policy movements, there are the effects of various rural programs from government, nongovernmental organizations, corporations, and other involved entities.

The market is far from the sole determiner of land use questions (e.g., Lee et al., 1995; Salano et al., 2001). Also, policy is somewhat perplexing in that the motives of land users, economic or otherwise, can transcend policy direction.

Without policy, land users do have more freedom to explore the ecological-agroecological interface to increase productivity and reduce risk. This may be why many of the examples of applied agroecology throughout this text originate in traditional societies untouched by policy.

# CHANGE

In Chapter 9, the different rotational patterns are discussed. Change can be more profound and long lasting. Certainly, there is an impetus to not fix that which is not broken. Examples exist where agrotechnologies adopted in ancient times persist to the present; ancient Roman landscape layouts are found in Europe (Paoletti, 2001), Inca terraces find use in Peru, and Asian rice paddies have existed in place for hundreds of years. These still address local needs but, if no longer in sync with socioeconomic conditions, they will submit to change.

Changes intended to last into the foreseeable future fall under the heading of redesign. This occurs when one or more agroecosystems are not in line with a changing economic situation, or when the land use situation and/or the advantages of introducing an entirely new system outweigh the costs or disadvantages associated with the change.

## *Redesign*

Gliessman (1998, p. 304) has discussed the different levels for increasing the sustainability of agriculture. These involve a change in agroecosystem design either minimally, through management inputs, or more consequentially, through an agrotechnological change. These are

1. efficiency gains that seek to increase the efficiency of the system with regard to inputs (e.g., changing the input nutrient balance),
2. substitution, where change involves substituting one type of input for another (e.g., biomass for fertilizer), and
3. a redesign where a single agroecosystem not in line with the changing economic or land use situation and/or the advantages of introducing an entirely new system (agrotechnology) outweighs the costs or disadvantages associated with the change.

The first two only slightly affect the broader landscape. Notable change at the landscape level starts with system redesign and proceeds to more substantial levels of modification. These are

4. cropping or rotational change (e.g., with the primary landscape species, one or more) that causes a shift in overall cropping patterns,
5. layout change brought about by access or other modifications as with the addition of roads or irrigation channels,
6. layout change that introduces more auxiliary systems including forest fragments or riparian systems,
7. countryside view, where there is a shift in the nature-agroecosystem interface (as part of biodiversity or cultural agroecology), or
8. complete landscape redesign that brings about a total shift in agroecological emphasis. Examples are a change from slash-and-burn to permanent cultivation or moving from single cropping to an inclusive taungya.

## *Economic*

Under an economic heading comes the continuation of landscape activities to maintain needed outputs or an income stream during the period of change. Subsistence farmers rely almost totally on staple outputs and, if land is intensely used, there is less opportunity for major change. Commercial forestry operations, those that supply wood processing centers, require a steady stream of timber and cannot tolerate much change. Commercial farms have more flexibility for change, but are still dependent on cash receipts.

Changing to more tree-based agriculture, although more income or greater yields accrue in the future, may be uneconomic only because of the time frame involved. Once established, these systems can be quite beneficial within the overall farm landscape. There are numerous examples, e.g., forest gardens, forest enrichment, or feed systems, that exist only because they are in place and producing.

A large leap into the unknown, despite the advantages of a comprehensive approach, is a major barrier to change. These considerations can make a series of small interventions (either plot by plot or small changes across the landscape) more appealing than a total landscape design.

## *Vegetative*

A number of changes result from socioeconomic need. An efficiency change is to select a plant species or variety that can use the site in question with little or no modification. Although hundreds (or at the extreme thousands) of varieties exist for common staple crops, this approach is seldom taken. The exception is where (1) numerous varieties are in widespread use, (2) differing growth attributes (DPCs) are understood, and (3) people accept varietal differences. This normally only occurs where the crop is indigenous to the region or has been used for hundreds, if not thousands, of years. Potatoes in the Andes, rice in Africa and Asia, and sweet potatoes in the Pacific islands are a few examples from an extensive list.

Site modifications can be vegetative. Draining a swamp by planting high-transpiration trees is an example. More common and less dramatic are those cases, discussed in this text, where smaller adjustments, e.g., temperature, are accomplished through agrotechnological or landscape change.

## *Physical*

As a result of a cropping change, the site may be modified to accommodate the preferred crops. This can involve physical modification in the land. Making swamps (rice paddies) where needed is a well-recognized example; draining swamps through channels is another. Other cases involve microcatchments and/or absorption zones that allow trees to be established where rainfall patterns are an obstacle. Terraces permit the use of marginal land for intensive agriculture.

## *Prioritizing*

Change is easy to discuss, but often difficult to undertake. Land users seldom have the farm resources available, i.e., labor or capital, to undertake a number of system changes, and therefore land users must establish some intervention criteria.

Ordering can also be important, as the introduction of one agrotechnology may make another less appealing. Alley cropping is more easily adopted if there is a lack of trees and wood products (Cooper et al., 1996), but may be less interesting if firewood is no longer needed.

Another landscape intervention influence is the coordination of agrotechnologies with regard to economic, environmental, and/or ethnographic needs. A full landscape evaluation may involve the simultaneous use of these criteria in multiple-criteria decision making.

Addressing environmental problems is another prioritizing criterion, with the most pressing environmental needs being initially addressed. This may start with those that affect staple or cash crops. Erosion control may be foremost, but other problems such as fertilizer runoff can be included under this heading. Other needs are lower on the priority list.

## LANDSCAPE AESTHETICS

As a quality-of-life issue, landscape aesthetics crosses various landscape boundaries and offers a number of intangible gains. Aesthetics can translate into financial gains through tourism or increased property values. As a topic, this is difficult to discuss as beauty is in the eye of the beholder.

### Elements

There have been attempts to explain the elements of aesthetics. Litton et al. (1974) list a high degree of unity (distinctness of a region), variety (changing patterns and textures), and vividness (contrasts in color and shading). Within this broader framework, a pleasing landscape unit also should have

1. boundary definition,
2. some unique feature (waterfalls, pinnacles, snow-capped mountains) or landform (lakes, valleys, etc.),
3. characteristic climate manifested through distinctive vegetation, patterns, and textures, and
4. culturally revealing land use patterns (field layout, cropping patterns, etc.).

These elements are provided more as food for thought than as definitive guidelines for embellishing a landscape.

### Types

Aesthetics comes in two forms: the landscape as a garden and/or the larger landscape as a visually pleasing entity. A farm or plantation need not become a botanical garden, but considerable latitude exists to use selected biodiversity to achieve visual distinction as well as fulfilling key agroecological objectives. This can have a cumulative effect, where as each plot or agroecosystem is made visually pleasing, the landscape itself is made more agreeable.

## The Garden Landscape

The addition of flowering, colorful, or texturally pleasing plants are a positive attraction. Wildflowers can be an addition to natural strips, blossoming hedge varieties can replace or augment standard varieties without loss of agroecological purpose, and fragrant flowers along agrarian roads can enrich the rural experience.

This should not be underrated. The number of aesthetic (ornamental) plants on farm holdings is indicative of the importance placed on them.

## The Rural Vista

Attractive landscapes are often epitomized as those with dramatic features (snow-covered peaks serve well, as do ocean vistas). Even without such striking features, much can be done to improve the local scenery. This requires using elevation and intersystem contours, vegetative textures, broken lines, and promoting natural features.

> *Contours*—Landscape can be made more pleasing by reducing the straight, angular look and adding contours and smooth lines. The opposite may hold true where angular plots contrast with rolling hills.
> *Texture*—A pleasing result is obtained by mixing textures. Blocks of regular pines can be made more visibly pleasing if the uniform texture is interrupted by deciduous species.
> *Broken impact*—Open spaces serve a purpose, as do natural areas. The best option may be a mix of the two. For example, a few lush parkland trees in an irregular placement can enhance open prairies and fields.
> *Key features*—Part of any landscape are natural features such as rivers, rock outcrops, cliff faces, and land-water interfaces. These should not be hidden, but emphasized through vegetative placement.

These suggestions represent only a few thoughts about achieving a visually pleasing landscape. This is very subjective topic, and one of the few art forms that engages all the senses.

# Chapter 13

# Cultural Motifs

Through their land use practices, people delineate an intimate association with nature, where elements of culture are outwardly manifested throughout the landscape. Cultural agroecology has many aspects, some smaller in scale and less visible. Others are visually apparent across a wide swath of countryside. This is the landscape cultural motif, where landscapes can be culturally demarcated by

1. the use of plots,
2. the type and form of agrotechnologies contained,
3. a preference for specific types of ecosystems and plant-plant interactions, and
4. economic criteria.

The easiest to observe, and one of the more common types, is a landscape comprising set agroecosystems in clearly defined plots. These are the one-plot one-agroecosystem or one-plot one-agrotechnology models discussed in the earlier chapters.

These agroecosystems constitute some of the many landscape variations that are encountered. In contrast, some individuals and groups have discarded the notion of prearranged agroecosystems and set plots in favor of more free-form, fuzzy-plot landscapes. There are numerous alternatives between these extremes.

Other options derive from what may be termed academic emphasis. This has a number of facets, one of which involves the role that agriculture and forestry or agroforestry play within the landscape. Another is based on the type of ecology professed (i.e., simple to complex).

Together with the economic factors discussed in the previous chapters, these determine the landscape motif. In most cases, the result is a repeating pattern, a sort of cultural stamp put upon the land that crosses various holdings. This originates from the views, perceptions, and aspirations of the participants but, because cultures share views and perceptions, they carry across a society and imprint the land with the tenets espoused.

## PLOT DOMINANT

In Chapter 1, a dichotomy was raised between the plot and the agroecosystem. This is important, as some regard the plot as being fundamental to any holding, superseding land characteristics (e.g., topographical, area microclimate, soil type, etc.) as a means to apportion agroecosystems across the landscape. Blocks are generally square or rectangular; other shapes are a rarity.

Although other views exist, a strong plot landscape is often associated with the one-plot one-agrotechnology model. A plot-dominant landscape is shown in Figure 13.1. This contrasts with the terrain-dominant landscape of Figure 13.2.

### Reasons

The motives for a strong plot landscape can include established (and hard to change) ownership patterns, locational factors (e.g., a strong village structure or high-output systems near major roads), and managerial factors (e.g., the economies of scale associated with large single tracts that cross topographical divisions). These reasons are often underscored by a cultural proclivity to skirt nature.

Strong plots do not always ignore landscape features, and some deference to nature may force layout decisions. For example, in areas of moderate topography, plots may dominate within moderate slope parameters, but not transgress topographical boundaries (e.g., active streams or steep hillsides).

Despite the clear scenes in Figures 13.1 and 13.2, it can be visually difficult to ascertain the relative strength of the plot as an individual or cultural tenet. Heavy stone walls or thick hedgerows may remain simply because they are too difficult to remove (as with change, see Chapter 12). A lack of fencing may only indicate that free ranging animals are not a problem. Where clear ecosystem boundaries are not evident, the placement of agroecosystems with a high regard to to-

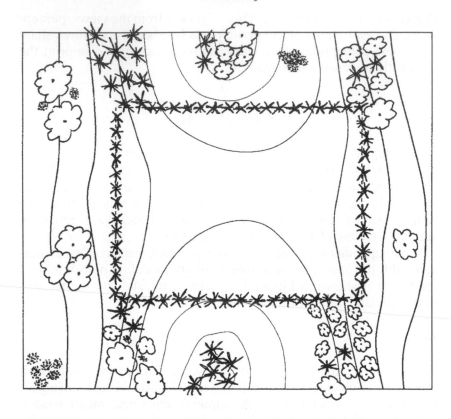

FIGURE 13.1. A plot-dominant motif means the plot has a regular shape regard-
less of the topography or other site characteristics.

pography can be an indication that plots are not an overriding consideration.

## Variations

The variations include plot size, content (as defined by the full design package), relative location, and the use of auxiliary systems. It is
possible to encompass more than one agroecosystem within a single
plot. This one-plot multiecosystem model indicates that the contained ecosystems share at least one common objective. The typical

FIGURE 13.2. A terrain-dominant landscape means the plot conforms to some attribute of the site. In this case, the plot area conforms to the slope gradient.

examples involve grazing, where an unshaded pasture may be coupled with an animal-sheltering ecosystem (see Photo 1.1).

Other variations are multiplot single-agroecosystem variants. The most visible examples employ dead fencing (as opposed to live hedgerows) that is not ecologically intrusive. An example is a large pasture subdivided into smaller plots.

A plot-dominant landscape can, but may not always, lead to a lesser environmental situation. Problems occur where agroecological principles are neglected and potential ecological gains are lost. For example, there can be a tendency to force crops to fit the site through applying inputs (fertilizers, irrigation, etc.), rather than using the finesse associated with matching the crop variety or plant-plant dynamics with the site characteristics. More often than not, problems will occur with large, single-ecosystem plots where the agroecosystem contained is not a good fit for the full area enveloped. Through subsequent agroecological weaknesses, this lack of fit can expose a site to the forces that promote soil, microclimatic, and ultimately, productive degradation.

If properly used, a strong plot approach can be as environmentally friendly as any other form. The key is in choosing contained agroecosystems that can arrest any immediate and long-term environmental threats.

## TERRAIN DOMINANT

The terrain-dominant model views landscape characteristics as more important than the plot. Here the agroecosystems are placed and configured in such a way that they conform to topography (e.g., slope), soil moisture, soil type, site microclimate, or other ecological characteristics of a site (see Photo 13.1). Plots often exist in this model, but they are secondary to the overlying terrain model. The viable manifestation is a fuzzy farm lacking straight boundaries, which may even lack clearly defined agroecosystems.

In pure form, crops (clone, cultivar, and/or species) are closely matched with soils, soil moisture, elevations, and host of other site characteristics. This type of placement with a clear boundary is shown in Figure 13.2. Often the boundary may be more transient, but this only reinforces the nonplot emphasis.

There are variations that reinforce this view. To take full advantage of terrain, a more drought-resistant species may be located higher on a hill. Better yet, a range of species may progress up a hillside as contour strips, thereby gaining all the productive and ecological advantages from the immediate and elevation-linked microsite (including soil, soil moisture, and temperature gradient). This can be a very ecologically favorable approach that sanctions a wide range of environmental gains.

A large area containing a single agroecosystem is usually not part of a terrain approach, but offers some opportunities for exploiting terrain differences. Other means, e.g., microsite arrangement or rotational patterns, can help fully utilize site differences and overcome

PHOTO 13.1. This example of a terrain-dominant landscape shows the agricultural plots placed on an area of lesser slope. This example is from the Andean mountains of central Chile.

any inherent ecological weaknesses, especially if the topography and other microsite differences are minute.

Although a mix of plot- and agroecosystem-dominant landscapes are possible, the tendency is to espouse one view or the other. A mixed system is an unusual case, derived from competing influences. As an alternative, semiplot or semiecosystem examples are discussed in the following section.

## AGROTECHNOLOGICAL REPRESENTATION

Within the sphere of cultural agroecology, many look at the agro-ecological landscape as a series of set agrotechnologies that are deployed with minimal alteration from the suggested prototype. This may not always be the case. Many individuals and cultures have moderated, transcended, or not subscribed to this view, instead looking at a landscape with less agrotechnological input.

The views range from a divisional landscape based on set agro-technologies (see Figure 13.3) to a landscape based on reformulated agrotechnologies. At the far end of this continuum is a landscape with only vestiges of, or completely lacking any, agrotechnology input.

Mixed landscapes (set, modified, or nonagrotechnological) are possible, but since this is not a uniform view, they may be limited to situations where a new agrotechnology or crop has been recently introduced and the modification process has not begun in earnest. Agrotechnological representation is independent of plot or terrain ranking.

Whatever form is adopted, there is still a need for agrotechnological trappings, i.e., DPCs, DAPs, a crude design package, etc. Lacking the set agrotechnology as a foundation, the ensuing landscapes can be more complex to install, but equally viable by any economic, ecological, or productive measure.

### Divisional Landscapes

A landscape comprising set, predetermined agrotechnologies is a divisional landscape. Chapter 3 explains the intricacies of the agrotechnologies, Chapters 4 and 5 present the different types, and subsequent chapters discuss dedicated use. These are only one building

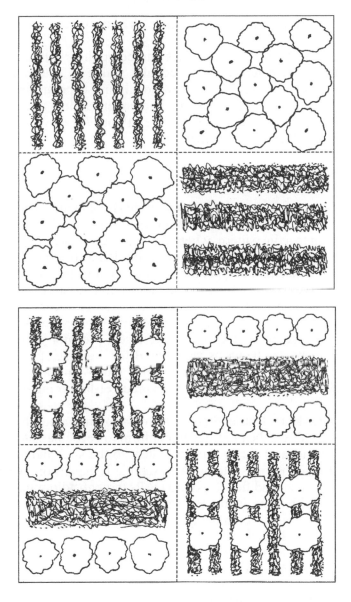

FIGURE 13.3. Comparing opposite corners of each four-plot landscape, trees and crops are shown in separate ecosystems (top) or in integrated form (bottom). Also evident is the use of simple ecology (the monocultures on top) or more complex bicultures (bottom).

block in landscapes where land users pick and choose among the alternatives offered.

This use of the agrotechnologies can be applied with a one-plot one-agroecosystem model or be equally viable in a terrain-based landscape. The advantages of a divisional landscape are more as a knowledge nucleus, where research and understanding are encapsulated into neat, easy to retrieve categories. Photo 7.1 shows a plot-oriented, set-agrotechnology model.

## Formulation Landscapes

A set agrotechnology, the essence of divisional agroecology, uses unmodified agrotechnologies in an evolved form. A divisional landscape is akin to a tile pattern, where the user cuts colored tiles to form a picture or pattern. Formulation agrotechnologies mean that the user can modify, to a small degree, the color content of the tiles.

This involves taking a set agrotechnology and modifying the internal design to change the mix of DAPs. The agrotechnologies are starting points, rather than end results. A small change might involve a hedgerow system where, to improve insect predator-prey properties, a multispecies hedge is substituted for a single species. This change might be carried further by also employing a cover crop within the crop row to supplement the predator-harboring properties of the hedge.

In an interdependent landscape, formulation has a number of advantages. DAPs can be modified to conform to particular site characteristics (the idea of fitting the ecosystem to the site) and/or for the benefit of neighboring systems. An example is expansion, where the agroecosystem is modified to extend DAPs through adjoining SIZ enlargement. With formulation, the set agrotechnologies are employed as starting points or recommended for use where systems are to undergo active modification.

## Placement Landscapes

The third viewpoint, the placement landscape, totally eschews the use of agrotechnologies, instead relying on other principles to formulate the landscape. The DPCs are chief among these. There are two variations of a placement landscape: semiplot and nonplot.

Because of the free-form nature of the landscape, they are most often found within a terrain-dominant approach. A plot approach is possible, but within the confines of the plot, the soil and land characteristics are the location criteria for vegetative placement.

## *Semiplot*

With a semiplot or semiecosystem landscape, there exist discernable areas (in this case, not always in blocks) where recognizable agrotechnologies (either set or formulated) are interspersed with the overall landscape. Other plants, mainly perennials, can be superimposed on the underlying principal-mode systems. These additions can constitute a single agrotechnology (where complementarity underlies the entire system, including any plant additions) or be a separate system (where soil or other natural features are best accommodated with different agroecosystems). The latter can be as small as a single tree located on a rock outcropping.

Lacking the underpinnings provided by the agrotechnologies, some other user guidelines are needed to steer the process. This is based on two tenets:

1. A perennial species is located where it grows best as long as it does not substantially interfere with more valuable land uses.
2. The amount (population) of any desirable species is proportional to its value, but in keeping with the first tenet.

Studies have shown similar guidelines in use. For example, den Biggelaar (1996, p. 69) found that species are chosen by

1. whether they damage the soil or associated crops,
2. where they fit within the farming system (near or away from crops),
3. where they grow best, and
4. their potential functions or uses (food, fiber, fencing, etc.).

Other studies document landscapes where plant placement is more dependent on site conditions and espouse a form of agriculture outside of the formal agrotechnologies (e.g., Padoch and de Jong, 1987).

*Nonplot*

Going even further, the notion of an agrotechnological underpinning may be partially or totally abandoned. An individual may instead opt to design a plot or terrain landscape where blocks are discernible, but the placement, size, and shape are directed more by topographical, soil, and climatic considerations than any need for order or uniformity. This design may use heavily modified, complex agrotechnologies, none of which will resemble the agrotechnology of origin (based on DAPs), or the basis may be the free-form placement of individual plants (using DPCs).

Facilitative plants may be purposely introduced. In doing so, each plant (productive and/or facilitative) is placed where it grows best without interfering with a more valuable species. Again, this rule is used without any agrotechnological underpinning.

In addition to a site-dictated spatial irregularity among productive species, nonplot design can have a temporal aspect where natural vegetation can infringe on crop blocks. The portion of biodiversity that is judged to have sufficient value and complementarity is allowed to remain. These are part of the natural forces at work (the occurring agroecosystem) and, if land users accommodate these forces, what evolves may mimic natural ecosystems.

The agroforest landscape of traditional South Pacific islands has an agrobiodiverse and rule-based system (e.g., Raynor, 1992). Other nonplot landscapes may be similarly agroforested, but in interspersed plots of single or mixed-crop species. In either case, these agroecosystems can have a transitional interface (lacking clear boundaries) or may utilize buffer species, buffer systems, or fuzzy makeup.

Although density, diversity, and disarray may offer a compelling structure, this is not the only underlying concept. In less dense and diverse ecosystems, DPCs and DAPs are the principles upon which a loose, biostructured landscape is formulated.

## ACADEMIC CLASSIFICATIONS

Other culturally based classifications influence or even determine the overall form of a landscape. For instructional purposes, universities designate them as colleges of study. Lacking a better term, they are used here: (1) agriculture, (2) forestry, and (3) agroforestry.

There are other subdivisions based on the degree of ecological complexity: (1) the monocultural perspective, (2) bicultural or tri-cultural agroecology built upon an understanding of individual and unlike plant-plant relationships, and (3) complex agroecology using ecosystem dynamics as the key driving force.

## Colleges

What occurs in practice may mirror the academic disciplines found in university systems. Thus, agroecosystems may be classified as to whether they originate in a college (or department) of agriculture or forestry (see Figure 13.4). A more recent addition, agroforestry, has not taken hold as a full academic discipline, but constitutes another view.

In the college paradigm, forestry and agriculture are in disparate plots with little or no functional overlap. The opposing model is agroforestry, where trees and crops are integrated within the landscape. This paradigm is separate from plot- or terrain-dominant approaches and independent of agrotechnological representation.

Although an indirect influence, college can be a powerful force. An example is farm forestry versus agroforestry, where the number of trees on a farm landscape may be the same, but their placement and use are entirely different.

This is more pervasive as a landscape influence than academic function suggests. Many government and international agencies are organized around the college archetype, where agronomists and foresters do not share the same institutional resources. This dichotomy permeates the policies and counsel of these groups. The college dichotomy has become a key landscape variable in commercial enterprises, is well established in many regions, and is spreading as advice is disseminated.

## Academic Ecological

Other academic variations are based on different forms of ecology, and these gain in-field influence through the perceptions and academic training that underlie agriculture and forestry. There are three divisions in agroecology:

1. The agroecological variants of complex natural ecology
2. Classic bicultural or tricultural agroecology
3. Simple agroecology

The first ecological form is based on natural ecosystems and dynamics that transcend simple plant-plant complementarity. This deals mostly with natural ecosystems where biodensity, biodiversity, and biodisarray are adopted for use with planned and managed systems; biodiverse mimicked systems are included.

The two variations are (1) highly biodiverse productive systems, such as agroforests, or, on the landscape level, (2) interspersed and interactive areas of natural vegetation. The key is reliance upon the natural dynamics associated with a fully functioning ecosystem. As a separate topic, this theme is developed in Chapter 11.

There are less biodiverse forms of agroecology, where the focus is on exploiting plant-plant complementarity. The landscape that results is bicultural or based upon various one-on-one complementarity relationships.

The third form is based on the simplest ecological systems, the relationships between niche-similar plants (i.e., monocultural systems). Although not as ecologically rewarding as other versions, this is a variation of agroecology where rotational and intersystem placement influences come to the fore.

The use of these versions is based upon land users' perceptions and knowledge of how to handle associated dynamics. Mixed landscapes are possible, such as the blend of monocultural rice, mixed-species gardens, and agroforests common throughout Southeast Asia. Intermingled ecology (i.e., where all forms of ecological complexity are present) comes into play

1. where this is within the comfort zone of land users (traditional and cultural acceptable land use),
2. where specific crops are assigned an ecological category based on local knowledge of these crops, and
3. where land users know how to blend agroecosystems to achieve the best economic and ecological outcome.

This is not a strong classification and, since the groupings are far from exclusive, they encompass much variation. Still, it provides a

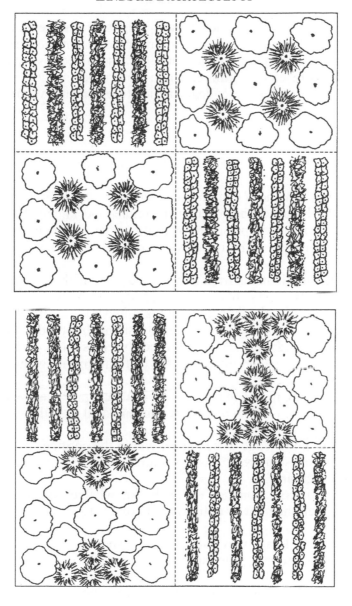

FIGURE 13.4. Two four-plot landscapes. At the top is a divisional landscape showing nearly identical ecosystems on opposite corners. At the bottom is a formulation landscape, where similar ecosystems (opposite corners) vary in design in response to a site, management, or other need.

high degree of cultural agroecological insight. This system of classification is autonomous in plot- or terrain-dominant landscapes, in agrotechnological representation, and, because mixed systems (agronomic and forestry) can still remain in their respective colleges, it is also autonomous in the college (agriculture forestry versus agroforestry) dichotomy.

## *MOTIF*

A motif is a theme or pattern that repeats across the landscape. Motifs are the broad expansion of perceptions and knowledge into a practical farm landscape and through the underlying beliefs, principles, and concepts of individuals and cultures (Wojtkowski, 2002, p. 326).

In part, a motif stems from having land users with similar dietary preferences, risk aversion, ritual needs, and societal concerns (Bellon, 1996). In league with topography, climate, and tradition, these dictate cropping selections, sites, and agroecosystems. Imposed upon these are the other landscape social influences (such as those discussed in this chapter). Together, they form the cultural imprint or motif that underlies a landscape.

There is more to it than cultural, societal, and practical needs suggest. This imprint is also part of an informational base that:

1. allows knowledge to be gained within the confines of a single agroecosystem with fewer variations and less complexity to consider, and
2. allows users to function within this knowledge base without being overwhelmed by options or alternatives.

Although a landscape may vary in design, species, etc., a more constricted set of cropping alternatives produces a narrower, but deeper, knowledge base. This can be a more effective vehicle in deriving an ecologically and economically efficient landscape than a wider, but less deep, base.

Among the advantages of this system-confined approach is that the training and experience needed to function effectively is lessened. A disadvantage is that the solutions are sought only within one motif, whereas other motifs may offer better alternatives. For example, the use of set agrotechnologies within a simple ecological emphasis produces a series of monocultural plots. Monocultures are more easily understood as research is conducted in relative simplicity. The disadvantage is that this offers fewer ecological options and gains than within other academic or ecological realms.

## Motif Variations

The variations in landscape motif produced from the two perceptions, academic and agrotechnological, can be quite profound, where each permutation produces another distinct landscape type. These are briefly discussed in this section.

The idea of a pure landscape type may seem out of place in real-world situations and thus thorough scrutiny may seem fruitless. However, some exist, some can be theorized, and all are encompassed in the scope of human thought. On this basis, discussion is merited.

In the real world, deriving the underlying motif compels the use of imagination, requires the intuition to distinguish differences through meager visual clues, and, as part of an end goal, obliges one to appreciate the underlying cultural values and the socioeconomic imperatives. In an unadorned form, the examples in this chapter are based upon visual clues.

Not all the cases presented in the text have corresponding illustrations, and not all permutations found in the real world can be illustrated. Enough are supplied to provide an overview as to what elements underscore a culture-based and/or socioeconomic-based system of landscape classification.

### Divisional Variations

From the two academic themes, ecology and college, and through the use of set agrotechnologies, a number of variations are produced. Using agronomy, forestry, or agroforestry as college forms and simple, plant-plant, and complex as ecological variations, nine pure variations exist.

Mixed forestry and agronomy landscapes (with trees in one plot, crops in another) are possible (as shown in Figure 13.3) (see Photo 13.2). These form another set of variations that are not directly discussed in this section.

PHOTO 13.2. A landscape exhibiting the cultural motifs shown in Figure 13.4 (bottom) and Figure 13.5. Rather than being abstractions, these patterns are encountered in the real world. This example is from Kenya.

Note that, by some definitions, agroforestry requires two or more interacting species (e.g., Méndez, 2001). This definition does not preclude the species from interacting at some distance (e.g., adjoining blocks); hence, simple agroforestry combinations exist. Strictly interpreted, it might be argued that this landscape type actually overlaps with, and is part of, a mixed landscape.

> *Agronomy/complex ecology*—This is a landscape clearly demarcated by ecosystems where each contains complex mixes of agronomic, short-rotation crops. The agrotechnologies can be mixed pasture, shrub gardens, or other agronomically classified, but species-biodiverse systems.
> *Agronomy/plant-plant ecology*—The landscape is based on intercrops in set plots. A maize-bean or maize-bean-squash landscape is an example of what might transpire with this approach.

*Agronomy/simple ecology*—This is a common landscape type composed of monocultural plots of short-duration species. Many regions in developed countries show this academic imprint. The Midwestern United States, lacking trees, is a pure form of this landscape type.

*Forestry/complex ecology*—This is composed of plots containing mixed forestry species such that natural dynamics come to the fore. A high-intensity forestry system with woodland blocks of different mixes and treatments is an example. This can be topography influenced.

*Forestry/plant-plant ecology*—Although great benefits have been reported from multispecies plantations (Kelty, 1992), landscape-wide examples of plant-plant (forestry biculture) ecosystems are lacking. The idea behind a full landscape design is to pair individual species, e.g., oak and pine, on the basis of interspecies complementarity or to match the system, through secondary species, to the site characteristics.

*Forestry/simple ecology*—This is more common in forestry plantation landscapes. Normally, there would be only one agrotechnology and one species, but plots of different age groupings also qualify. Ecologically, more can be accomplished by varying variety or species, but this runs counter to a trend toward more use of clones.

*Agroforestry/complex ecology*—This type of landscape may epitomize ecology, offering the greatest range of ecological benefits and design options (see Figure 13.5). Unfortunately, this type of landscape is far from common. It can be found on South Pacific islands (Raynor, 1992). Where found, these landscapes may be semi- or nonplot, where the divisions are harder to discern.

*Agroforestry/plant-plant ecology*—This landscape variation is uncommon, but can be observed. The customary use is a taungya landscape where, in pure form, different stages of extended taungyas constitute separate bicultural systems.

*Agroforestry/simple ecology*—This landscape type is occasionally observed. Plots of trees are interspersed with plots of crops. This is the same as a mixed agronomic or forestry variation with monocultures, but where the blocks are relatively small and tree-crop interactions are the norm.

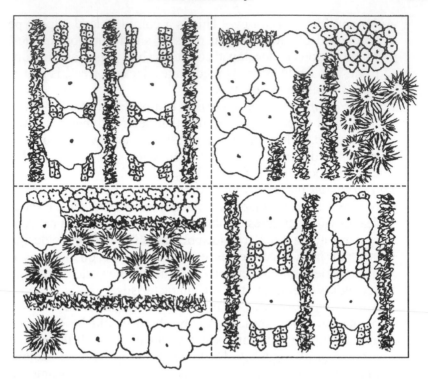

FIGURE 13.5. A system-mixed landscape based upon complex agroforestry. The upper left and lower right plots are divisional; the others (upper right and lower left) are formulation based. The latter landscape forms are pictured in Photo 13.2.

## Formulation Variations

The formulation variations closely follow the nine subdivisions of a divisional landscape. The difference is in having more internal agroecosystem variation, more opportunity for agroecosystems modeled on internal dynamics, and more opportunity for interecosystem effects.

## Placement Variations

As with formulation, the variations derive from the nine divisional variations. As this is akin to a painting with an infinite number of variations, the degree of complexity can be difficult to qualify. This does not mean these landscape are unmanageable, only that they require procedures for doing so.

*Temporal Variations*

Landscapes can be fairly static affairs with little year-to-year change in each plot, such as sole cropping or a rotation sequence utilizing the same static agrotechnology. The alternative is based on a more intricate planned sequence that progresses from one agrotechnological type to another, as with an extended taungya sequence.

*Economic Variations*

Economics can be the key guide in many, if not most, planned and managed landscapes. This is not an exception with motif types. Two different intersystem criteria direct the landscape process: (1) economic agroecosystem classification (with internally or externally sustained systems) and (2) the use of economic orientation (land use intensity with high- and low-yielding plots or a more balanced plot input-output approach). Their use can be an offshoot of topography, a cultural norm, or both.

The use of economic interrelationships produces another set of permutations, which subdivide motif into (1) externally funded, (2) internally supported, and (3) mutually sustained agroecosystems. This is added to the already discussed use of the agrotechnologies and the two academic variations; college (i.e., agronomy, forestry, and agroforestry) and ecological (i.e., simple, plant-plant, and complex). With these, there are 27 variations ($3 \times 3 \times 3$), only a sample of which are discussed here.

>   *Agronomic/simple/externally funded*—For a divisional landscape
>       composed of agronomic/simple ecology, an additional varia-
>       tion uses one of the three economic types. In the most com-
>       mon landscapes, set agroecosystems are externally funded. An
>       example is a series of plot-based monocultures, each receiv-
>       ing chemical inputs.
>   *Agronomic/simple/self-sustained*—Another option, self-sustain-
>       ing ecosystems, is a possibility and examples exist. The most
>       expedient way to accomplish this using simple systems is

through rotations (see the case study in Chapter 8, a medieval landscape).

*Agronomic/simple/mutually supported*—An agronomic landscape with simple ecology is a bit more difficult to implement with any degree of ecological accomplishment, but can be mutually sustained through the clever use of plot content, placement, and rotation.

*Forestry/simple/self-sustained*—A divisional landscape composed of monoculture trees that are ecologically self-sustained is the norm in large-scale plantation forestry. The long rotations of each component keep nutrient outflows within acceptable limits, while the unplanned subsidiary vegetation is conducive to favorable predator-prey and other ecological dynamics.

*Forestry/simple/externally funded*—An ecologically simple forestry landscape that is externally funded exists with many tree crop systems. Although outside a strict definition of forestry, the examples are tree crop plantations (oil palms, rubber trees, etc.), where greater nutrient demands compel additional inputs.

*Forestry/simple/mutually supporting*—Although no clear examples have been described of a forestry/simple ecosystem that is mutually sustaining, the possibilities are real. More than likely, they find use in regions with varying topography where species are arranged in a mutually supporting manner.

*Agroforestry/simple/externally funded*—As agroforestry systems are not simple, this may seem like a contradiction in terms. However, what evolves from these three parameters is a series of monocultural agroecosystems, some tree based, others having crops. The agroforestry element is obtained through interaction (and placement) of the trees and crops.

## Economic Orientation

As a landscape criterion, orientation, whether revenue or cost, is another factor that is interrelated with economic-based classifications. These may be less visual than those described, but still quite discernible. Because of this, these variations are not illustrated. The different expressions of orientation may be more important as a land-scape-expanding tool than as a separate landscape grouping.

## *Motif Derivation*

The cases of pure landscape types evolve from the academic-agrotechnological continuum. Although these explain some of the underlying landscape motives and directions, they do not go far enough in interpreting what is happening and how the process occurs.

### *Ground Rules*

Any landscape is formulated around certain rules. Work by den Biggelaar (1996, p. 70) shows how farmers position individual species. Other cultures use similar systems and, based upon this, some basic ground rules can be proposed. The landscape version of these rules are as follows:

1. Agroecosystems and individual plants are located where they grow best, provided this does not eclipse more productive land uses.
2. The amount of a plant species in the landscape is proportional to its value, except where constrained by soil and risk factors.

These rules guide the expansion of a farm landscape. These can be used with all agrotechnological variations including divisional and formulation-derived landscapes. These rules are especially useful in guiding nonplot layouts. The general idea is that the primary land-scape species (the most valued plant, staple or market) has first choice of the area occupied. Beyond this, crops or trees are placed with regard to their contribution in the overall landscape. This can be based solely on market value, which can generate an agroecosystem-independent landscape, or with regard to agroecological value, which can produce a more biodiverse, interdependent appearance.

### *Progressions*

Within the context of the ground rules, a number of cultural or viewpoint progressions exist that guide a productivity-based land-scape design into a final form. The starting point is often a staple or high-value crop on the best site. The best site is relative, as for some, it may be fertile, well-drained bottomland. For others on more marginal lands, the best site may be a low-fertility hillside.

This is based on a ceteris paribus premise, as other factors may override this scenario. As presented in previous chapters, this can include prior use (where a land user may not want to remove an existing and producing system) or policy factors (where, due to price supports or other government programs, a normally low-value crop may be the more valuable choice).

There is an interlinkage between progression and economic orientation. The starting points are normally systems that are revenue oriented, often externally funded. As more agroecosystems are added, they are often less revenue oriented. When the first systems are proposed, the notion of mutual support is not often foremost. However, as more productive units are added and/or knowledge is gained, this can come into play.

## LANDSCAPE DERIVATION

The design or derivation of a landscape can follow a subscribed pattern and/or the ground rules. In pure form, design proceeds along set motif lines and is an optimization procedure. Four approaches to landscape design are

1. exploited natural ecosystem,
2. independent agroecosystem,
3. a plot-centrific approach, and
4. community net present value (community NPV).

Each has its own reasons for being. A fifth, overlay, is used to modify the initial process to increase the ecological and productive capacities.

One must remember that landscape design is seldom an immediate process. It occurs over an extended period (measured in years, if not decades) and, as the process progresses and individual ecosystems are modified, knowledge is added as to best use and placement of species and systems. During this period views change. The result will not always yield a pure motif. This is reflected in the methods, especially overlay, that can occur as a later development. The starting points and progressions can be plot or terrain dominant, within a college, or employ one or more versions of agroecology.

The relationships between these approaches are shown here, where each can be used alone or supplemented by overlay.

1. Natural (exploited) ecosystem ————————→ Overlay
2. Independent agroecosystem ————————→ Overlay
3. Plot-centrific ————→Principal-mode ————→ Overlay
                              ↘ Auxiliary ——→ Overlay
4. Community NPV ————————————————→ Overlay

## *Natural (Exploited) Ecosystem*

In any natural ecosystem subject to exploitation, there are patterns of use. Examples are (1) logging with its access and feeder (skid) roads and landings where logs are put on trucks, (2) subsistance gathering in forests with its patterns of activity, or (3) grazing in grasslands or nomadic lifestyles with seasonal animal movement.

## *Independent Agroecosystem*

The basis of this landscape type are the independent agroecosystems, either externally funded or ecologically self-contained. Normally, they are divisional landscapes with set agrotechnologies (the one-plot one-agrotechnology model), but some degree of agrotechnological flexibility is possible.

From the starting point, a landscape evolves based on a series of externally funded and/or self-contained agroecosystems. They can be forest or crop based. Each system is placed where it grows best (using ground rules) with little regard to intersystem influences.

The externally funded version occurs commonly on commercial farm landscapes. The self-contained version is more common in commercial forestry enterprises. A mixed system, externally funded and self-contained, is found with farm forestry where separate blocks of trees are located near or between cropping systems.

## Plot-Centrific Approach

Outside the domain of the independent agroecosystem is a land-scape composed of ecologically interacting systems. This requires planning such that, as the landscape evolves from the starting point, each system complements another. The center point (plot and/or agroecosystem) is usually the highest quality site on the holding.

Again, this can be based on the one-plot one-agroecosystem model. More likely, it may involve a terrain-dominant landscape.

A single plot is usually selected and the landscape design radiates from it. This is done using DAPs, which are the basis for each subsequent ecosystem placement.

A plot-centrific approach has two subbranches or options, (1) principal-mode or (2) auxiliary supporting landscape. As these are agro-technologically based, they are found only with divisional or formulation landscapes.

### Principal Mode

In pure form, this is a landscape composed of principal-mode systems that, through DAPs and SIZs, exert the needed agroecological influence on neighboring agroecosystems. Ideally, this is accomplished with a formulation landscape by modifying the ecological and economic properties of individual agrotechnologies. More options are introduced with agroecosystem expansion or augmentation, where a DAP of one system is carried across to a nearby system.

### Auxiliary

This subbranch of the DAP-centrific approach does not rely heavily upon the DAPs of individual principal-mode systems, but uses auxiliary systems to provide or increase the needed SIZs. Landscape structures such as cut-and-carry systems can be part of this landscape design.

As with a pure principal-mode design, expansion and augmentation are an option, but they may prove less viable than adjusting the size, shape, and location of agroecosystems to take advantage of needed SIZs.

## Community NPV

The use of community NPV to manage a biodiversity-based agro-ecosystem was presented in Chapter 11. This approach also serves as a basis for landscape design. For community NPV, no one agro-ecosystem serves as a starting point, but some or all agroecosystems may share this role.

Each ecosystem is looked at in terms of present and future value (using the NPV to relate future value to the present), including auxiliary and natural ecosystems. Each is placed and formulated such that the value is maximized across the entire landscape. This is more intricate than the plot-centrific approach and may find use more with ecosystem replacement or modification.

## Postdesign Overlay

The overlay approach is a subbranch of neither natural ecosystem, independent, plot-centrific, or community NPV design. It is based upon the placement of individual plants or modestly used auxiliary systems. For cost reasons, it is used mainly with woody perennials, although annual plants in narrowly defined roles, e.g., repelling insects, are possible.

Overlay can be part of a grand plan or, more likely, used to augment the existing structure providing any missing and beneficial dynamics. This is not exclusive to either an independent ecosystem or centrific approach. Overlay is valuable in that any ecological shortcomings or limitations that result from prior use may be overcome through overlay.

Overlay may also occur in natural landscapes subject to exploitation. It can be shade trees planted in natural grasslands or trees planted as part of forest enrichment in a natural forest ecosystem.

The overlay landscape may be identical in appearance to a self-contained or DAP-based landscape. For example, a parkland system based on a set agrotechnology can be the same as that obtained through tree overlay on a crop plot.

The temporal sequence, no matter how intricate, can be part of overlay. The difference is that the additional elements remain in place throughout the sequence, often emerging unchanged at the end of the full temporal sequence. This is found in tropical West Africa where some tree species remain uncut during the slash-and-burn phase, are part of the cropping phase, and are part of the subsequent extended fallow.

# Chapter 14

# Conclusion

The previous chapters have looked at a wide array of factors that influence the layout, use, and outcome of an agroecological landscape. To be fully effective, the agroecological landscape constitutes a mix of principles, practices, and concepts, often with a large dose of compromise.

## *Infrastructure*

Physical infrastructure needs, roads, buildings, water channels, etc., can set the tone for a landscape, or be only a relatively minor influence. In more commercial holdings, these are fixed in place, not subject to modification, and a possible impediment to change. In slash-and-burn agriculture, they may be only a transient hindrance to a redesign.

Along the same lines are land ownership patterns that can protect or destroy the ecological integrity of a site. The tragedy of the commons stems from a lack of ownership. Layout patterns that run counter to topography, water flows, and resulting natural processes are examples of situations where the ownership pattern is a detriment to the full exercise of agroecological principles.

## *Spatial Layout*

For any given economic land use entity, the climate and topography dictate what type of systems are employed. Shelterbelts grace many windswept, rolling countrysides, those where wind drying limits crop and tree growth. Hillside agriculture almost always requires some form of contour planting to achieve the best outcome. These are only a few examples in a long list of agroecological applications.

Prior use has been mentioned, and the imprint on the land can be hard to erase. Often one must work around or work to modify existing placements. This includes economic barriers to change.

It is possible to broaden agroecological appeal by changing the size, content, and timing of the contained ecosystems (e.g., a change to a one-plot multiecosystem style). Landscape principles are maximized when the individual systems are small and there exists ample opportunity to exploit all interecosystem effects. This need not compel a major shift in layout; only plot content through space and time. Other gains come with other approaches, such as increasing biodiversity, involved rotational dynamics, the use of complementarity, DPCs, and/or DAPs.

## Temporal Considerations

Landscapes are seldom stable, but are subject to a host of factors that force change. Outside of rotational dynamics, factors can include market forces (where fluctuating crop-selling prices or new markets alter the productive mix), and the need to conform to population pressures, and/or to respond to technology innovations (including agroecological inputs). Any change must work around an existing infrastructure, accommodate new entities, and function within land right constraints.

Within this broader context, there are changing plot-plot or agroecosystem-on-agroecosystem relationships that can be used to advantage. They take a number of forms.

1. Temporal-facilitative, where each cropping sequence sets the nutrient or pest management stage for the next, i.e., leaves the site more suitable for ensuing crops.
2. Spatial-facilitative, where, in changing crop situations, the facilitative benefits of having one system located near another are preserved.
3. Spatial complementarity, where adjoining systems have plant-plant complementarity between the component species. Keeping these relationships reduces interface distance and promotes more efficient land use.

When incorporating these relationships, options are opened within the temporal landscape. The taungyas are examples of long-term strategies that, once implemented over a large area, can provide agroecological benefits.

## The Agrotechnologies

Set or prepackaged agrotechnologies offer a number of advantages. They limit the complexity of any land use problem addressed and provide a clear unit for knowledge gathering and extension. In providing prototypes from which to work, the agrotechnologies provide an unambiguous assortment of viable options for immediate use or for further refinement.

Eschewing these, there is no reason that a land user cannot work with a more free-form fuzzy landscape. The concepts of complementarity, desirable plant characteristics, and multipurpose plants spawn a range of landscape types. Landscapes not based upon agrotechnological usage (the nonagrotechnological, free-form landscape) are found in some societies. Outside the agrotechnological context, these landscape types require that each land user have a good knowledge of plants and their relationship to soils and climate.

For the uninitiated, landscape design without an underlying agrotechnological structure can be a daunting task. Because of this, the nonagrotechnological landscape may be reserved for those who better understand the layout parameters for the more exotic designs or have a set of ground rules from which to work.

## Biodiversity

Landscapes can be formulated using biodiverse agrotechnologies, biodiversity at the fringe of plots, and/or a more free-form internal approach employing ecosystems enhanced through selected biodiversity. For this landscape type, the yield restrictions are somewhat apparent, the agroecological gains fairly clear.

It is known that nature operates successfully on a set of basic principles (including density, diversity, and disarray), and these have been usurped by foresters and farmers to promote certain activities. Outside of impacted forests and agroforests and throughout the wider landscape, it remains a challenge to transfer these principles to a crop-diverse, prosperous, and productive landscape.

## The Socioeconomic Outcome

In viewing all landscape options, one must keep a steady eye on the objectives and outcomes. This perspective is what differentiates the agroecological landscape from what nature provides. Nature offers stress control without cost, while individuals must invest time and resources to achieve their specific rendering. Unfortunately, nature is not very accommodating when it comes to the production of staple crops. The further removed from that which nature intended, the greater the management investment needed.

Outcome is measured through economic criteria. Use of these depends upon the situation. The more agroecological the landscape, the greater need for evaluations that follow along agroecological lines. Quality of life is on this list, as are business decisions regarding allocation of scarce, and not so scarce, resources across the various means of production. Risk is also part of any economic determination that, because of the increased number of agroecosystem and landscape options offered, comes more to the fore in agroecology.

## Cultural Agroecology

Agroecological landscapes take many forms and are the result of a vast array of influences. Cropping systems are directed, in part, through tradition, religion, and culture. For example, eating cattle and pigs are religious taboos, while a prohibition on consuming bugs is merely an offshoot of a squeamish culture. Component plants contribute in varying ways; one is the production of prestigious items that figure prominently in cultural expression, e.g., yams and breadfruit in the South Pacific (Raynor, 1992).

There are other examples where cultural input has a bearing on field practice. The nonharvest option, where only a portion of the harvestable output is removed, has cultural overtones. Some groups regard this as a wasteful practice; others take advantage of the agroecological options offered.

Cultural agroecology takes many forms. Some relate to established in-field cropping needs and practices (as briefly described previously); others are landscape-wide, putting an indelible stamp or mark on the landscape.

## Cultural Motif

Landscape design involves more than physical and socioeconomic determinants. By residing in the same region, land users are swayed by similar climatic forces, soil types, land-holding dimensions, site restrictions on crops, and risk threats to these crops. Land users may also have similar dietary preferences, socioeconomic needs, and constraints on the amount of labor and financial resources available. Also within this mix are limits on knowledge that may confine groups to a set array of farming and/or forestry practices.

The different cultural influences can be enigmatic. To the uninformed, the motif may be hidden, only visible through unique and subtle clues. For others, it is concealed by perspective.

What makes this process more mystifying is that individuals, operating within or outside cultural parameters, put their own mark on the land. They can be progressive as they try innovative techniques. Others, steeped in tradition, may resist change and maintain some legacy of the past. Either viewpoint is valid. In combination, these enrich existing practice by promoting a diversity of thought in agroecological theory and providing limits to applied agroecology. All this can lead to advances in land use practice.

The importance of cultural motif lies in being able to suggest cropping changes that conform to the norms of the local society and are more readily adopted. Motif allows policy to be formulated, not only on the basis on institutional perceptions, but on cultural traditions and individual efforts of land users. From the agroecological perspective, motif can be used to separate meaningful and unique agroecological contributions from land use practices with little agroecological grounding.

## Ecological Gains

Land degradation is a gradual process, as it first loses those intangible, nonessential, and less-noticed benefits (e.g., populations of songbirds and clean, clear water). Eventually, after what may be a long downward trend, the productive potential of the land may be lost.

This process can be reversed, future directions changed, more accommodation with nature made, and human aspirations, traditions, and values advanced. A single overwhelming agroecological improvement may overcome a major obstacle to agroecological health. The opposite and more likely scenario is one of numerous small ecological gains, each contributing to restoring the whole, with some improvements yielding larger gains than others. How far land users are willing to go in restoration or improvement depends upon knowledge, aspirations, and the resources committed.

### General Landscape Guidelines

Given the multiplicity of ideas presented throughout this text, more confusion than clarification may be the end product. It is advantageous to offer a wide spectrum of ideas but, given the diversity of conventions found throughout the world, putting them in some coherent framework is a bit more difficult. However, there are some loose guidelines that seem appropriate. They serve to start, not end, the thought process.

Landscape agroecology can emphasize different approaches. Among the options are

- the use of plant-plant complementarity packaged as principal-mode agrotechnologies and distributed according to agroecological principles;
- the use of the plant-plant interface, where a landscape is formulated based on complementarity and DPCs between unlike species that are placed accordingly;
- ecosystem dynamics, where ecological gains accrue because, in highly biodiverse systems, the ecological interactions exceed the gains that are possible through individual plant-plant associations alone;
- rotational dynamics, where gains stem from the planned replacement of existing vegetation;
- the interaction between individual species (through DPCs) and ecosystems (through DAPs) with placement determined by these properties; and

- the use of DAPs to formulate an array of mutually reinforcing agroecosystems, each formulated for individual sites and the needs of neighboring systems.

The use of these approaches is not exclusive. The better landscapes employ a mix of approaches with one dominant theme, such as the landscape design alternatives, e.g., plot-centrific, overlay, etc. Equally, the mechanisms and systems utilized do not focus on a single natural stress, but address them in order of relevance and degree of risk brought into play.

## *Landscape Agroecology*

The introduction to this text states that effective agroecology is a knowledge-driven process, one that does not stem solely from the productive potential of high-input agriculture, but also rests on a multiplicity of agroecological concepts and conventions. These concepts can be acquired from those who have overcome unique local obstacles or have made headway toward achieving production in what may be termed nonconventional agricultural and forestry settings (deserts, high altitudes, etc.). Studies from ecology, agronomy, forestry, agroforestry, and related land use and social disciplines also contribute.

The ideas sanctioned in the preceding chapters should help in plumbing the profundity of practices and views that constitute landscape agroecology. At the minimum, the goal is a landscape that serves as a productive unit, enriching the human experience.

More fitting is a landscape designed to use ecological dynamics to promote or achieve productive purposes, reinforcing the socioeconomic and cultural values of the inhabitants, all while allowing natural flora and fauna to thrive. It is only through agroecology that the highest bar can be reached, and this is accomplished only by employing all available tools. Such is the relevance of an agroecologically derived landscape.

# References

Agelet, A., Bonet, M.A., and Vallès, J. (2000). Homegardens and their role as a main source of medicinal plants in mountain regions of Catalonia (Iberian Peninsula). *Economic Botany* 54(3):295-309.

Alomar, O., Goula, M., and Albajes, R. (2002). Colonization of tomato fields by predatory mirid bugs (Hemiptera: *Heteroptera*) in northern Spain. *Agriculture Ecosystems and Environment* 89:105-115.

Altieri, M.A. and Trujillo, J. (1987). The agroecology of corn production in Tlaxcala, Mexico. *Human Ecology* 15(2):189-220.

Aman, R. (1998). Rare and wild fruits of Peninsular Malaysia and their potential uses. In: Nair, M.N.B., Sahri, M.H., and Asharri, Z. (Eds.), *Sustainable Management of Non-Wood Forest Products* (pp. 204-211). Proceedings of an International Workshop—Universiti Putra Malaysia, October 14-17, 1997, Serdang: Universiti Putra Malaysia Press, 327 pp.

Anderson, M.K. (1999). The fire, pruning, and coppice management of temperate ecosystems for basketry material by California Indian tribes. *Human Ecology* 27(1):79-85.

Ashley, M.D. (1986). *Agroforestry in Haiti*. Orono, ME: University of Maine, 69 pp.

Ataroff, M. and Rada, F. (2000). Deforestation impact on water dynamics in a Venezuelan Andean cloud forest. *Ambio* 29(7):440-444.

Attenborough, D., Hogarth, J., and Saito, S. (2000). Japan's Secret Garden. *NOVA* Broadcast No. 2716, PBS air date: December 19.

Ayuk, E.T. (1997). Adoption of agroforestry technology: The case of live hedges in the central plateau of Burkina Faso. *Agricultural Systems* 54:189-206.

Bäckman, J.-P.C. and Tianen, J. (2002). Habitat quality of field margins in a Finnish farmland area for bumblebees (Hymenoptera: *Bombus* and *Psithyrus*). *Agriculture Ecosystems and Environment* 89:53-68.

Bainbridge, D.A. (1988). The oaks: A neglected multiuse treecrop. In: Allen, P. and van Dussen, D. (Eds.), *Global Perspectives on Agroecology and Sustainable Agricultural Systems: Proceedings of the Sixth International Scientific Conference of the International Federation of Organic Agriculture Movements* (pp. 557-601). Santa Cruz, CA: The Ecology Program, 730 pp.

Banda, A.Z., Meghembe, J.A., Ngugi, D.N., and Chome, V.A. (1994). Effect of intercropping maize and closely spaced *Leucaena* hedgerows on soil conservation and maize yield on a steep slope at Ntcheu, Malawi. *Agroforestry Systems* 27:12-22.

Banful, B., Dzietror, A., Ofori, I., and Hemeng, O.B. (2000). Yield of plantain alley cropping with *Leucaena leucocephala* and *Flemingia macrophylla* in Kumasi, Ghana. *Agroforestry Systems* 49:189-199.

Baskin, Y. (1994). Ecologists dare ask: How much does diversity matter. *Science* 264:202-203.

Beckerman, S. (1983). Barí swidden garden: Crop segregation patterns. *Human Ecology* 11(1):85-101.

Beckerman, S. (1984). A note on ringed fields. *Human Ecology* 12(2):203-206.

Beer, J. (1987). Advantages, disadvantages and desirable characteristics of shade trees for coffee, cacao and tea. *Agroforestry Systems* 5:3-13.

Bellon, M.R. (1996). The dynamics of crop intraspecific diversity: A conceptional framework at the farmer level. *Economic Botany* 50(1):26-39.

Belsky, A.J. (1992). Effects of trees on nutrient quality of understory gramineous forage in tropical savannas. *Tropical Grassland* 26:12-20.

Belsky, A.J. and Canham, C.D. (1994). Forest gaps and isolated savanna trees. *Bio-Science* 44(2):77-84.

Bettiol, W. (1999). Effectiveness of cow's milk against zucchini squash powdery mildew *(Sphaerotheca fuliginea)* in greenhouse conditions. *Crop Protection* 18(8):489-492.

Boffa, J.-M. (2000). West African agroforestry parklands: Keys to conservation and sustainable management. *Unasylva* 51:11-17.

Boffa, J.-M., Yaméogo, G., Nikiéma, P., and Toanda, J.-B. (1996). What future for the shea tree? *Agroforestry Today* 8(4):5-9.

Boster, J. (1983). A comparison of the diversity of Jivaroan gardens with that of the tropical forests. *Human Ecology* 11(1):47-68.

Bradburd, D. (1982). Volatility of animal wealth among Southwest Asian pastoralists. *Human Ecology* 10(1):85-106.

Bragança, M., De Souza, O., and Zanuncio, J.C. (1998). Environmental heterogeneity as a strategy for pest management in *Eucalyptus* plantations. *Forest Ecology and Management* 102:9-12.

Brenner, A.J. (1996). Microclimatic modifications in agroforestry. In: Ong, C.K. and Huxley, P. (Eds.), *Tree-Crop Interactions: A Physiological Approach* (pp. 159-188). Wallingford, UK: CAB International, 386 pp.

Buresh, R.J. (1999). Phosphorus management in tropical agroforestry: Current knowledge and research challenges. *Agroforestry Forum* 9(4):61-65.

Caborn, J.M. (1965). *Shelterbelts and Windbreaks.* London: Faber and Faber, 287 pp.

Caramori, P.H., Androcioli Filho, A., and Leal, A.C. (1996). Coffee shade with *Minosa scabrella,* Benth. for frost protection in southern Brazil. *Agroforestry Systems* 33:205-214.

Carter, J. (1996). Alley farming: Have resource-poor farmers benefited? *Agroforestry Today* 8(2):5-7.

Carucci, R. (2000). Trees outside the forest: An essential tool for desertification control in the Sahel. *Unasylva* 51(200):18-24.

Chandrashekara, U.M. and Sankar, S. (1998). Ecology and management of sacred groves in Kerala, India. *Forest Ecology and Management* 112(1-2):165-177.

Charles, D. (2002). Fields of dreams. *New Scientist* January 5:25-27.

Christanty, L., Kimmins, J.P., and Mailly, D. (1997). "Without bamboo, the land dies": A conceptual model of the biogeochemical role of bamboo in an Indonesian agroforestry system. *Forest Ecology and Management* 91:83-91.

Cieugh, H.A., Miller, J.M., and Böhm, M. (1998). Direct mechanical effects of wind on crops. *Agroforestry Systems* 41:85-112.

Coates, K.D. and Burton, P.J. (1997). A gap-based approach for development of silvicultural systems to address ecosystem management objectives. *Forest Ecology and Management* 99:337-354.

Colbach, N., Lucas, P., Cavelier, N., and Cavelier, A. (1997). Influences of cropping system on short eyespot in winter wheat. *Crop Protection* 16(5):415-422.

Cooper, P.J.M., Leakey, R.R.B., Rao, M.R., and Reynolds, L. (1996). Agroforestry and the mitigation of land degradation in the humid and sub-humid tropics of Africa. *Experimental Agriculture* 32:235-290.

Davis, J.H.C., Roman, A., and Garcia, S. (1987). The effects of plant arrangement and density in intercropped beans *(Phaseolus vulgaris)* and maize, II. Comparison of relay intercropping and simultaneous planting. *Field Crops Research* 16:117-126.

Davis, J.H.C., Woolley, J.N., and Moreno, R.A. (1986). Multiple cropping with legumes and starchy roots. In: Francis, C.A. (Ed.), *Multiple Cropping Systems* (pp. 133-160). New York: Macmillan, 383 pp.

Day, S. (2001). Fight the blight. *New Scientist* September 1(2306):36-39.

den Biggelaar, C. (1996). *Farmer Experimentation and Innovation: A Case Study of Knowledge Generation Process in Agroforestry Systems in Rwanda.* Community Forestry Case Study Series 12, Rome: FAO, 123 pp.

D'Hondt-Defrancq, M. (1993). Nematodes and agroforestry. *Agroforestry Today* 5(2):5-8.

Donovan, T.M. and Flather, C.H. (2002). Relationships among North American songbird trends, habitat fragmentation, and landscape occupancy. *Ecological Applications* 12(2):346-374.

Ehui, S.K. (1992). Population density, soil erosion, and profitability of alternative land-use systems in the tropics: An example in southwestern Nigeria. In: Sullivan, G.H., Huke, S.M., and Fox, J.M. (Eds.), *Financial and Economic Analysis of Agroforestry Systems: Proceedings of a Workshop Held in Honolulu, Hawaii, USA, July 1991* (pp. 95-106). Paia, HI: Nitrogen Fixing Tree Association, 312 pp.

Ellis, B.A. and Bradley, F.M. (1996). *The Organic Gardener's Handbook of Natural Insect and Disease Control.* Emmaus, PA: Rodale Press, 534 pp.

Escalante, E.E. (1995). Coffee and agroforestry in Venezuela. *Agroforestry Today* 7(3-4):5-7.

Evans, J. (1992). *Plantation Forestry in the Tropics*, Second Edition. Oxford: Clarendon Press, 403 pp.

Faizool, S. and Ramjohn, R.K. (1995). *Agroforestry in the Caribbean.* Santiago, Chile: Oficina Regional de la FAO para American Latina y El Caribe, 35 pp.

FAO (1976). *Harvesting Man-Made Forests in Developing Countries: A Manual on Techniques, Roads, Production and Costs.* Rome: FAO, 197 pp.

FAO (1994). *Prácticas Agroforestales en el Departmento de Potosí—Bolivia: Documemto de Trabajo No 1.* FAO/ HOLANDA/ CDF, Proyecto "Desarrollo Forestal Comunal el Altiplano Boliviano." Bolivia: Potosí, 134 pp.

Farley, R.A. and Fitter, A.H. (1999). Temporal and spatial variation in soil resources in a deciduous woodland. *Journal of Ecology* 87:688-696.

Fox, T.R. (2000). Sustained productivity in intensively managed forest plantations. *Forest Ecology and Management* 138:187-202.

Gamero, E.M., Lok, R., and Somarriba, E. (1996). Análisis agroecológico de huertos caseros traditionales en Nicaragua. *Agroforestería en las Americas* 3(11-12):36-40.

Garrity, D.P. (1996). Tree-soil-crop interactions on slopes. In: Ong, C.K. and Huxley, P. (Eds.), *Tree-Crop Interactions: A Physiological Approach* (pp. 299-318). Wallington, UK: CAB International, 386 pp.

Gaye, S. (1987). Glaciers of the desert. *Ambio* 16(6):351-356.

Geertz, C. (1972). The wet and the dry: Traditional irrigation in Bali and Morocco. *Human Ecology* 1(1):23-39.

Gerritsma, W. and Wessel, M. (1997). Oil palm: Domestification achieved? *Netherlands Journal of Agricultural Sciences* 45:463-475.

Glausiusz, J. (2003). A green renaissance for the Sahel. *Discover* 24(1):13.

Gliessman, S.R. (1998). *Agroecology: Ecological Processes in Sustainable Agriculture.* Ann Arbor, MI: Ann Arbor Press, 357 pp.

Godoy, R. and Bennett, C.P.A. (1991). The economics of monocropping and intercropping by smallholders: The case of coconuts in Indonesia. *Human Ecology* 19(1):83-98.

Gongfu, Z. (1982). The mulberry dike-fish pond complex: A Chinese ecosystem of land-water interactions on the Pearl River Delta. *Human Ecology* 10(2):191-202.

Gras, N.S.B. (1940). *A History of Agriculture in Europe and Americas.* New York: F.S. Crofts and Co., 496 pp.

Gupta, G.N., Mutha, S., and Limba, N.K. (2000). Growth of *Albizia lebbeck* on micro-catchments in the Indian arid zone. *International Tree Crops Journal* 10: 193-202.

Hansen, S.W. (1996). Is agricultural sustainability a useful concept? *Agricultural Systems* 50:117-143.

Hansen, S.W. and Jones, J.W. (1996). A system framework for characterizing farm sustainability. *Agricultural Systems* 51:185-201.

Harvey, C.A. (2000). Windbreaks enhance seed dispersal into agricultural landscapes in Monteverde, Costa Rica. *Ecological Applications* 10(1):155-173.

Harvey, C.A., Haber, W.A., Solano, R., and Mejías, F. (1999). Arboles remanentes en potreros de Costa Rica: ¿Herramientas para la conservación? *Agroforestería en las Américas* 9(24):19-22.

Henrich, J. (1997). Market incorporation, agricultural change, and sustainability among the Machiguenga Indians of the Peruvian Amazon. *Human Ecology* 25(2):319-323.

Herzog, F. and Oetmann, A. (2001). Communities of interest and agroecosystem restoration: Streuobst in Europe. In: Flora, C. (Ed.), *Interactions Between Agroecosystems and Rural Communities* (pp. 85-102). Boca Raton, FL: CRC Press, 273 pp.

Hitimana, N. and McKinley, R.G. (1998). The effects of intercropping on phytophagous pests: A review. *Agroforestry Forum* 9(2):9-11.

Holmgren, P., Masakha, E.J., and Sjöholm, H. (1994). Not all African land is being degraded: A recent survey of trees on farms in Kenya reveals rapidly increasing forest resources. *Ambio* 23(7):390-395.

Hughes, J.D. (1994). *Pan's Travail: Environmental Problems of the Ancient Greeks and Romans.* Baltimore, MD: John Hopkins University Press.

Hulugalle, N.R. and Ezumah, H.C. (1991). Effects of cassava-based cropping systems on physio-chemical properties of soil and earthworm casts in a tropical alfisol. *Agriculture, Ecosystems and Environment* 25:55-63.

IACPA (1998). *Integrated Farming; Agricultural Research into Practice,* A report from the Integrated Arable Crop Production Alliance for Farmers, Agronomists and Advisors. London: Ministry of Agriculture, Fisheries and Food, MAFF Publications, 16 pp.

IITA (1976-1982). *International Institute of Tropical Agriculture Annual Reports.* Ibadan, Nigeria.

Jackson, W. (2002). Natural ecosystems agriculture: A truly radical alternative. *Agriculture, Ecosystems and Environment* 88:111-117.

Jain, S.K. (2000). Human aspects of plant diversity. *Economic Botany* 54(4):459-470.

Jensen, M. (1993). Soil conditions, vegetation structure and biomass of a Javanese homegarden. *Agroforestry Systems* 24:171-186.

Jensens, J.W. (1990). Landscape development scenarios for planning and implementing agroforestry: A case study in the semi-arid lands of eastern Kenya. In: Budd, W.W., Duchhart, I., Hardestry, L.H., and Steiner, F. (Eds.), *Planning for Agroforestry* (pp. 267-292). Amsterdam: Elsevier, 338 pp.

Joffee, R., Rambal, S., and Ratte, J.P. (1999). The dehesa system of southern Spain and Portugal as a natural ecosystem mimic. *Agroforestry Systems* 45:57-79.

Johnson, A.W. (1972) Individuality and experimentation in traditional agriculture. *Human Ecology* 1(2):149-159.

Jones, C.A. and Kiniry, J.R. (1986). *CERES-Maize: A Model of Maize Growth and Development.* College Station, TX: Texas A&M University Press.

Jordan, V.W.L., Hutcheon, J.A., and Donaldson, G.V. (1997). The role of integrated arable production systems in reducing synthetic inputs. *Aspects of Applied Biology* 50:419-429.

Jordan, V.W.L., Hutcheon, J.A., Donaldson, G.V., and Farmer, D.P. (1997). Research into and development of integrated farming systems for less-intensive arable production: Experimental progress (1998-1994) and commercial implementation. *Agriculture, Ecosystems and Environment* 64:141-148.

Jordan, V.W.L., Hutcheon, J.A., Glen, D.M., and Farmer, D.P. (1996). *Technology Transfer of Integrated Farming Systems: The LIFE Project,* Third Edition. Bristol, UK: Integrated Arable Crops Research, 24 pp.

Josiah, S.J., Prestin, S., and Gullickson, D. (1999). Cost/benefit analysis of living snow fences in Minnesota. Paper presented at *The Sixth Conference on Agroforestry in North America,* Hot Springs, AK, June 12-16.

Kareiva, P. (1994). Diversity begets productivity. *Nature* 368:686-697.

Kass, D.C.L. and Somarriba, E. (1999). Traditional fallows in Latin America. *Agroforestry Systems* 47:13-36.

Kelty, M.J. (1992). Comparative productivity of monocultures and mixed-species stands. In: Kelty, M.J., Larson, B.C., and Oliver, C.D. (Eds.), *The Ecology and Silviculture of Mixed Species Forests* (pp. 125-141). The Netherlands: Kluwer Academic Publishers, 287 pp.

Kendall, L. (1980). Forest berries and mushrooms—An endangered resource? *Ambio* 11(5):241-247.

Kidundo, M. (1997). *Melia volkensii*—Propagating the tree of knowledge. *Agroforestry Today* 9(2):21-22.

Koech, E.K. and Whitbread, R. (2000). Disease incidence and severity on beans in alleys between leucaena hedgerows in Kenya. *Agroforestry Systems* 49:85-101.

Komar, B.M., Kumar, S.S., and Fisher, R.F. (1998). Intercropping teak with *Leucaena* increases tree growth and modifies soil characteristics. *Agroforestry Systems* 42:81-89.

Kormawa, P.M., Kamara, A.Y., Jutzi, J.C., and Sanginga, N. (1995). Economic evaluation of using mulch from multi-purpose trees in maize-based production systems in south-western Nigeria. *Experimental Agriculture* 35:101-109.

Kowsar, A. (1992). Desertification control through floodwater spreading in Iran. *Unsylva* 43(168):27-30.

Kühlmann, D.H.H. (1988). The sensitivity of coral reef to environmental pollution. *Ambio* 27(1):13-21.

Kumar, H.D. (1999). *Biodiversity and Sustainable Conservation.* Enfield, NH: Science Publishers, 409 pp.

Kvist, L.P. and Nebel, G. (2001). A review of Peruvian flood plains: Ecosystem, inhabitants and resource use. *Forest Ecology and Management* 150:3-26.

Lagerlöf, J., Goffre, B., and Vincent, C. (2002). The importance of field boundaries for earthworm (Lumbricidae) in the Swedish agricultural landscape. *Agriculture, Ecosystems and Environment* 89:91-103.

Lee, D.J., Tipton, T., and Pingsun Leung (1995). Modelling cropping decisions in a rural developing country: A multiple-objective programming approach. *Agricultural Systems* 49:101-112.

Lefroy, E.C., Hobbs, R.J., Connor, M.H., and Pate, J.S. (1999). What can agroforestry learn from natural ecosystems? *Agroforestry Systems* 45:423-436.

Linares, O.F. (1976). Garden farming in the American tropics. *Human Ecology* 4(4):331-349.

Liping, Z. (1991). Biological control of insect pests and plant diseases in agroforestry systems. In: Avery, M.E., Cannell, M.G.R., and Ong, C.K. (Eds.), *Biological Research for Asian Agroforestry* (pp. 73-87). Morrilton, AR: Winrock International; New Delhi: IBH Publishing Co., 285 pp.

Litton, R.B., Tetlow, R.J., Sorensen, J., and Beatty, R.A. (1974). *Water and Landscape: An Aesthetic Overview of the Role of Water in the Landscape*. New York: Water Information Center Inc., 314 pp.

Liyanage, M. de S. (1993). The role of MPTS in coconut-based farming systems in Sri Lanka. *Agroforestry Today* 5(3):7-8.

Long, C. (2001). Bring on the bugs. *Organic Gardening* 48(4):10.

Lovett, P.N. and Naq, N. (2000). Evidence for the anthropic selection of the sheanut tree *(Vitellaria paradoxa)*. *Agroforestry Systems* 48:273-288.

Ma, M., Tarmi, S., and Helenius, J. (2002). Revisiting the species-area relationship in a semi-natural habitat: Floral richness in agricultural buffer zones in Finland. *Agriculture, Ecosystems and Environment* 89:137-148.

MacDonald, K.I. (1998). Rationality, representation, and the risk mediating characteristics of a Karakoram mountain farming system. *Human Ecology* 26(2):287-317.

Mäder, P., Fliessbach, A., Dubois, D., Gunst, L., Fried, P., and Higgli, U. (2002). Soil fertility and biodiversity in organic farming. *Science* 296(5573):1694-1697.

Malcolm, J.R. (1994). Edge effects in central Amazonian forest fragments. *Ecology* 75(8):2438-2445.

McNeely, J.A. (1993). Economic incentives for conserving biodiversity: Lesson from Africa. *Ambio* 22(2-3):144-150.

Meir, A. and Tsoar, H. (1996). International borders and range ecology: The case of Bedouin transborder grazing. *Human Ecology* 24(1):39-64.

Melman, D.C.P. and Van Strien, A.J. (1993). Ditch banks as a conservation focus in an intensively exploited peat farm. In: Vos, C.C. and Opdam, P. (Eds.), *Landscape Ecology of a Stressed Environment* (pp. 123-141). London: Chapman & Hill, 310 pp.

Méndez, E. (2001). An assessment of tropical homegardens as examples of sustainable local agroforestry systems. In: Gliessman, S.R. (Ed.), *Agroecosystem Sustainability: Developing Practical Strategies* (pp. 61-64). New York: CRC Press, 210 pp.

Menzel, P. and D'Alvisio, F. (1998). *Man Eating Bugs: The Art and Science of Eating Insects*. Berkeley, CA: Ten Speed Press, 191 pp.

Mertz, O. (1998). Wild vegetables as potential new crops in farming systems of Serwak, Malaysia. In: Nair, M.N.B., Sahri, M.H., and Asharri, Z. (Eds.), *Sustainable Management of Non-Wood Forest Products* (pp. 70-83). Proceedings of an International Workshop—Universiti Putra Malaysia, October 14-17. Serdang: Universiti Putra Malaysia Press, 327 pp.

Michon, G. and de Foresta, H. (1997). Agroforests: Predomestication of forest trees or tree domestication of forest ecosystems. *Netherlands Journal of Agricultural Science* 45:451-462.

Milius, S. (2002). Petite pollinators. *Science News* 161(5):69-70.

Miller, D.R., Lin, J.D., and Lu, Z.N. (1991). Some effects of surrounding forest canopy architecture on the wind field in small clearings. *Forest Ecology and Management* 45:79-91.

Mingkui Cao, Shijun Ma, and Chunsu Han (1995). Potential productivity and human carrying capacity of a agro-ecosystem: An analysis of food production in China. *Agricultural Systems* 47:387-414.

Mohd Ali, A.R. and Raja Bariza, R.S. (2001). Intercropping rattan with rubber and other crops. *Unasylva* 52(205):9-11.

Moore, R.W. and Bird, P.R. (1997). Agroforestry systems in temperate Australia. In: Gordon, A.M. and Newman, S.M. (Eds.), *Temperate Agroforestry* (pp. 119-148). Wallingford, UK: CAB International, 269 pp.

Morales, H., Perfecto, I., and Ferguson, B. (2001). Traditional fertilization and its effect on corn insect populations in the Guatemalan highlands. *Agriculture, Ecosystems and Environment* 84:145-155.

"Much ado about nothing." (2002). *New Scientist* May 18(2343):44-47.

Nabham, G.P. and Sheridan, T.E. (1977). Living fencerows of the Rio San Miguel, Sonora, Mexico: Traditional technology for floodplain management. *Human Ecology* 5(2):97-111.

Naeem, S., Thompson, L.J., Lawler, S.P., Lawton, J.H., and Woodfin, R.M. (1994). Declining biodiversity can alter the performance of ecosystems. *Nature* 368: 735-737.

Nair, N.P.K. (1990). Classification of agroforestry systems. In: MacDicken, K.G. and Vergara, N.T. (Eds.), *Agroforestry: Classification and Management* (pp. 31-57). New York: John Wiley and Sons, 382 pp.

Nair, N.P.K. (1993). *An Introduction to Agroforestry.* Dordrecht, Netherlands: Kluwer Academic Publishers, 499 pp.

Nebel, G. (2001). Sustainable land-use in Peruvian flood plain forests: Options, planning and implementation. *Forest Ecology and Management* 150:187-198.

Nepal, S.K. and Weber, K.E. (1995). Prospects for coexistence: Wildlife and local peoples. *Ambio* 24(4):238-245.

Nevo, Y.D. (1991). *Pagans and Herders: A Re-Examination of the Negev Runoff Cultivation System of the Byzantine and Early Arab Period.* Jerusalem: Achva Press, 259 pp.

Ngulube, M.R. (1995). Indigenous fruit trees in southern Africa: The potential of *Uapaca kirkiana. Agroforestry Today* 7(3-4):17-18.

Nicholls, C.I. and Altieri, M.A. (2001). Manipulating plant biodiversity to enhance biological control of insect pests: A case study of a northern California organic vineyard. In: Gliessman, S.R. (Ed.), *Agroecosystem Sustainability: Developing Practical Strategies* (pp. 29-48). New York: CRC Press, 210 pp.

NRC (1996a). *Ecologically Based Pest Management*. Washington, DC: National Research Council, Board of Agriculture, 144 pp.

NRC (1996b). *Lost Crops of Africa*, Volume 1, *Grains*. Washington, DC: National Research Council, Board on Science and Technology for International Development, 393 pp.

Nyong, A.O. and Kanaroglou, P.S. (1999). Domestic water use in rural semiarid Africa: A case study of Katarko Village in northeastern Nigeria. *Human Ecology* 27(4):537-543.

O'Connor, R.J. and Shrubb, M. (1986). *Farming and Birds*. New York: Cambridge University Press, 290 pp.

Owiunji, I. and Plumptre, A.J. (1998). Bird communities in logged and unlogged compartments in Budongo Forests, Uganda. *Forest Ecology and Management* 108:115-126.

Padoch, C. and de Jong, W. (1987). Traditional agroforestry practices on native and ribereno farmers in the lowland Peruvian Amazon. In: Gholz, H.L. (Ed.), *Agroforestry: Realities, Possibilities and Potential* (pp. 179-194). Boston: Dordecht Publishers.

Paoletti, M.G. (2001). Biodiversity in agroecosystems and bioindicators of economic health. In: Shiyomi, M. and Koizumi, H. (Eds.), *Structure and Function in Agroecosystem Design and Management* (pp. 11-44). New York: CRC Press, 435 pp.

Pearce, F. (2001). Desert harvest. *New Scientist* 172(2314):44-17.

Perfecto, I., Rice, R.A., Greenberg, R., and Van der Voort, M.E. (1996). Shade coffee: A disappearing refuge for biodiversity. *BioScience* 46(8):598-608.

Pest control: A dynamite idea. (2001). *The Economist* June 2:82.

Peterson, J.T. (1981). Game, farming, and inter-ethnic relations in northeastern Luzon, Philippines. *Human Ecology* 9(1):1-22.

Platt, J.O., Caldwell, J.S., and Kok, L.T. (1999). Effects of buckwheat as a flowering border on populations of cucumber beetle and their natural enemies in cucumber and squash. *Crop Protection* 18(5):305-313.

Rachie, M.D. (1983). Intercropping tree legumes with annual crops. In: Huxley, P.A. (Ed.), *Plant Research in Agroforestry* (pp. 103-116). Nairobi, Kenya: ICRAF, 595 pp.

Rao, M.R. (1986). Cereals in multiple cropping. In: Francis, C.A. (Ed.), *Multiple Cropping Systems* (pp. 96-132). New York: Macmillan, 383 pp.

Rao, M.R. and Gacheru, E. (1998). Prospects of agroforestry for *Striga* management. *Agroforestry Forum* 9(2):22-26.

Raynor, W. (1992). Economic analysis of indigenous agroforestry: A case study on Pohnpei Island, Federated State of Micronesia. In: Sullivan, G.H., Huke, S.M., and Fox, J.M. (Eds.), *Financial and Economic Analysis of Agroforestry Systems:*

*Proceedings of a Workshop held in Honolulu, Hawaii, USA, July 1991* (pp. 243-258). Paia, HI: Nitrogen Fixing Tree Association, 312 pp.

Redford, K.H., Klein, B., and Murcia, C. (1992). Incorporation of game animals into small-scale agroforestry systems in the neotropics. In: Redford, K.H. and Padoch, C. (Eds.), *Conservation of Neotropical Forests* (pp. 334-346). New York: Columbia University Press, 475 pp.

Rice, R.A. and Greenburg, R. (2000). Cacao cultivation and the conservation of biological diversity. *Ambio* 29(3):167-173.

Rigby, D. and Cáceres, D. (2001). Organic farming and the sustainability of agricultural systems. *Agricultural Systems* 68:21-40.

Risch, S.J. and Carroll, C.R. (1982). Effect of a predaceous ant, *Solenopis geminata*, on arthropods in a tropical agroecosystem. *Ecology* 63:1979-1983.

Ritchie, J.T., Singh, U., Godwin, D., and Hunt, L. (1991). *A User's Guide to CERES-Maize.* Simulation Manual IFDC-SM-1, 5C, Second Edition. Boca Raton, FL: International Fertilizer Development Centre.

Rouw, A. de (1995). The fallow period as a weed-break in shifting cultivation (tropical wet forests). *Agriculture, Ecosystems and Environment* 54:31-43.

Ruc, J. (1990). Compounds from plants that regulate or participate in disease resistance. In: Chadwick, D.J. and March, J. (Eds.), *Bioactive Compounds from Plants* (pp. 213-224). New York: John Wiley and Sons, 242 pp.

Ruhigwa, B.A., Gichura, W.P., Spencer, D.S.C., and Swennen, R. (1994). Economic analysis of cut-and-carry and alley cropping systems of mulch production for plantains in south-eastern Nigeria. *Agroforestry Systems* 26:313-138.

Salam, A.M., Mohanakumaran, N., Jayachandran, B.K., Mammen, M.K., Sreekumar, D., and Babu, K.S. (1991). Thirty-one tree species support black pepper vines. *Agroforestry Today* 3(4):16.

Salano, C., León, H., Pérez, E., and Herrero, M. (2001). Who makes farming decisions, a study of Costa Rican dairy farmers. *Agricultural Systems* 67:153-179.

Sanchez, P.A. (1995). Science in agroforestry. *Agroforestry Systems* 30:5-55.

Sandström, K. (1998). Can forests "provide" water: Widespread myth or scientific reality. *Ambio* 27(2):132-138.

Schofield, N.J. (1992). Tree planting for dryland salinity control in Australia. *Agroforestry Systems* 20:1-23.

Schroth, G., Krass, V., Gasparotto, L., Duarte-Aguilar, J.A., and Vohland, K. (2000). Pests and diseases in agroforestry systems of the humid tropics. *Agroforestry Systems* 50:199-241.

Shrair, A.J. (2000). Agricultural intensity and its measure in frontier regions. *Agroforestry Systems* 49:301-318.

Sidibé, J.F., Scheuring, J.F., Tembely, D., Sidibé, M.M., Hofman, P., and Frigg, M. (1996). Baobab—Homegrown vitamin C for Africa. *Agroforestry Today* 8(2):13-15.

Simberoff, D. (1999). The role of science in the preservation of forest biodiversity. *Forest Ecology and Management* 115(2-3):101-111.

Sinclair, F. (1999). A general classification of agroforestry practices. *Agroforestry Systems* 46:161-180.

Siwatibau, A. (1984). Traditional environmental practices in the South Pacific— A case study of Fiji. *Ambio* 13(5-6):365-368.

Smith, N.J.H. (1996). Home gardens as a springboard for agroforestry development in Amazonia. *International Tree Crops Journal* 9:11-30.

Snyder, W.E. and Ives, A.R. (2001). Generalist predators disrupt biological control by a specialist parasitoid. *Ecology* 820(3):705-716.

Soluri, J. (2001). Altered landscapes and transformed livelihoods: Banana companies, Panama disease and rural communities on the north coast of Honduras, 1880-1950. In: Flora, C. (Ed.), *Interactions Between Agroecosystems and Rural Communities* (pp. 15-30). New York: CRC Press, 273 pp.

Ssekabembe, C.K., Hendelong, P.R., and Larson, M. (1997). Effect of hedgerow orientation on maize light interception and yield in black locust alleys. *International Tree Crops Journal* 9:109-118.

Stanton, M. and Young, T. (1999). Thorny relationships. *Natural History* 108(9): 28-31.

Stocks, A. (1983). Candoshi and cocamilla swiddens in eastern Peru. *Human Ecology* 11(1):69-84.

Stringer, W.C. and Alverson, D.R. (1994). Use of fire as a management tool in alfalfa production ecosystems. In: Cambell, K.L., Graham, W.D., and Bottcher, A.B. (Eds.), *Environmentally Sound Agriculture* (pp. 492-493). Proceedings of the Second Conference, April 20-22, Orlando, FL, 577 pp.

Styger, E., Kakotoarimanana, J.E.M., Rabcvohitra, R., and Fernandes, E.C.M. (1999). Indigenous fruits of Madagascar: Potential components of agroforestry systems in improved human nutrition and restored biological diversity. *Agroforestry Systems* 46:289-310.

Torquebiau, E. (1992). Are tropical agroforestry home gardens sustainable? *Agriculture, Ecosystems and Environment* 41:189-207.

Trawick, P.B. (2001). Successfully governing the commons: Principles of social organization in an Andean irrigation system. *Human Ecology* 29(1):1-26.

Tremblay, A., Mineau, P., and Stewart, R.K. (2001). Effect of bird predation on some pest insect populations in corn. *Agriculture, Ecosystems and Environment* 83:143-152.

Tscharntke, T., Steffan-Dewenter, I., Kruess, A., and Thies, C. (2002). Contribution of small habitat fragments to conservation of insect communities of grassland-cropland landscapes. *Ecological Applications* 12(2):354-363.

Tucker, C.M. (1999). Private versus common property forests: Forest conditions and tenure in a Honduran community. *Human Ecology* 27(2):201-210.

UN (2000). *Statistical Yearbook—1997*. New York: United Nations.

Unruh, J.D. (1990). Interactive increase of economic tree species in managed swidden-fallows of the Amazon. *Agroforestry Systems* 11:175-197.

Valderrábano, J. and Torrono, L. (2000). The potential for using goats to control *Genista scorpius* shrubs in European black pine stands. *Forest Ecology and Management* 126:377-383.

van der Valk, H.C., Niassy, A., and Bèye, A.B. (1999). Does grasshopper control create grasshopper problem?—Monitoring side-effects of fenitrothian applications in the western Sahel. *Crop Protection* 18(2):139-149.

Van Mele, P. and van Lenteren, J.C. (2002). Survey of current management practice in a mixed-ricefield landscape, Mekong Delta, Vietnam—Potential of habitat manipulation for improved control of citrus leafminer and citrus red mite. *Agriculture, Ecosystems and Environment* 88:35-48.

Vandermeer, J. (1989). *The Ecology of Intercropping*. Cambridge, UK: Cambridge University Press, 237 pp.

Vandermeer, J. (1995). The ecological basis of alternative agriculture. *Annual Review of Ecology and Systematics* 26:210-224.

Vasey, D.E. (1992). *An Ecological History of Agriculture 10,000 BC-AD 10,000*. Ames, IA: Iowa State University Press, 363 pp.

Velasco, A., Ibrahim, M., Kass, D., Jiménez, F., and Rivas Platero, G. (1999). Concentraciones de fósforo en suelas bajo sistema silvipastoril de *Acacia mangiun* con *Brachiaria humidicola*. *Agroforestería en las Américas* 6(23):45-47.

Versteeg, M.N., Amadji, F., Eteka, A., Gogan, A., and Koudokpon, V. (1998). Farmers' adoptability of mucuna fallowing and agroforestry technologies in the coastal savanna of Benin. *Agricultural Systems* 56:269-287.

Vickery, J., Carter, N., and Fuller, R.J. (2002). The potential value of managed field margins as forging habitats for farmland birds in the UK. *Agriculture, Ecosystems and Environment* 89:41-52.

Vieyra-Odilon, L. and Vibrams, H. (2001). Weeds as crops: The value of maize weeds in the valley of Toluca, Mexico. *Economic Botany* 55(3):426-443.

Waddell, E. (1975). How the Enga cope with frost: Responses to climate perturbation in the Central Highlands of New Guinea. *Human Ecology* 3(4):249-273.

Wang, H. (1994). Tea and trees: A good blend from China. *Agroforestry Today* 6(1):6-8.

Wells, G.W. (2002). *Biotechnical Streambank Protection: The Use of Plants to Stabilize Streambanks*. Lincoln, NE: National Agroforestry Center, Agroforestry Notes, Note 23, 4 pp.

Whitcomb, W.H. (1970). Woodpeckers in the ecology of southern hardwood borers. In: *Proceedings of the Tall Timbers Conference on Ecological Animal Control by Habitat Management* (pp. 309-315), Volume 2, February 26-28. Tallahassee, FL: Tall Timbers Research Station, 286 pp.

Wijesinghe, D.K. and Hutchings, M.J. (1999). The effects of environmental heterogeneity on the performance of *Glechoma hederacea:* The interactions between patch contrast and patch scale. *Journal of Ecology* 87:860-872.

Williams, T. (2002). America's largest weed. *Audubon* 104(1):24-31.

With, K.A., Pavuk, D.M., Worchuck, J.L., Oates, R.K., and Fisher, J.L. (2002). Threshold effects of landscape structures on biological control in agroecosystems. *Ecological Applications* 12(1):52-65.

Wojtkowski, P.A. (1993). Toward an understanding of tropical home gardens. *Agroforestry Systems* 24:215-222.

Wojtkowski, P.A. (1998). *The Theory and Practice of Agroforestry Design.* Enfield, NH: Science Publishers, 282 pp.

Wojtkowski, P.A. (2002). *Agroecological Perspecives in Agronomy, Forestry and Agroforestry.* Enfield, NH: Science Publishers, 375 pp.

Wojtkowski, P.A., Brister, G.H., and Cubbage, F.W. (1988). Using multiple objective programming to evaluate multi-participant agroforestry systems. *Agroforestry Systems* 7:185-195.

Yoon, C.K. (2000). Simple method found to increase crop yield vastly. *The New York Times* August 22:D1-D2.

Yoon, C.K. (2001). Something missing in fragile cloud forest: The clouds. *The New York Times* November 20:D4

Young, A. (1989a). *Agroforestry for Soil Conservation.* Wallingford, UK: CAB International, 275 pp.

Young, A. (1989b). The environmental basis of agroforestry. In: Reifsnyder, W.E. and Darnhofer, T.O. (Eds.), *Meteorology and Agroforestry* (pp. 29-48). Nairobi, Kenya: ICRAF, 546 pp.

Zhang Fend (1996). Influences of shelterbelts in Chifeng, Inner Mongolia. *Unasylva* 47(185):11-15.

Zimmerer, K.S. (1999). Overlapping patchwork of mountain agriculture in Peru and Bolivia toward a regional-gobal landscape model. *Human Ecology* 27(1): 135-165.

# Index

Page numbers followed by the letter "f" indicate figures; those followed by the letter "t" indicates tables; those followed by the letter "p" indicate photos.